Management for Professionals

For further volumes:
http://www.springer.com/series/10101

Peter Curwen • Jason Whalley

Fourth Generation Mobile Communication

The Path to Superfast Connectivity

 Springer

Peter Curwen
University of Strathclyde
Glasgow
United Kingdom

Jason Whalley
Newcastle Business School
Newcastle
United Kingdom

ISSN 2192-8096   ISSN 2192-810X (electronic)
ISBN 978-3-319-02209-3   ISBN 978-3-319-02210-9 (eBook)
DOI 10.1007/978-3-319-02210-9
Springer Cham Heidelberg New York Dordrecht London

Library of Congress Control Number: 2013954850

Printed on acid-free paper

Springer is part of Springer Science+Business Media (www.springer.com)

Preface

Roughly, 1 decade has passed since so-called 3G became established in a significant number of countries, and it is now available in virtually every country in the world—the case studies in the book illustrate where, and why, this is not the case. Hence, it is time to move on to consider the next generation of mobile technology, known universally as 4G—although, as the book notes, this term is incorrectly applied to what is currently available. The intervening technologies such as 3.5G are also covered, but our main concern here is to look to the future.

Technological advance is all about the speed at which data can be transmitted to mobile devices, primarily handsets. Unfortunately, as most readers will be aware, what is promised is very unlikely to be what you get due to the understandable tendency for operators to advertise meaningless maximum speeds. 3G can be a very disappointing experience even though it is much faster than what could be delivered by previous technologies and in general suffices for most purposes. The problem is that almost as soon as it becomes possible to do certain tasks that require a considerably faster downlink, the demand arises to complete more complex tasks that require that downlink to be significantly improved. Whereas email and fast access to websites were the ambitions a decade ago, the ambition today is to download huge video files, for example films, within a matter of seconds—or, to put it another way, to expect everything to be accessible almost instantly.

In general, transfer speeds via fixed wire are much faster than those available to mobile devices, and this of itself tends to promote a desire to do everything via a mobile link that can be done via a fixed-wire link. This ambition has yet to be fully realised, but (possibly well) before this decade is out the 500 kilobits per second (500 kbps) downlink commonly available in 2000 will have become 100 million bits per second (100 Mbps). Eventually, much learned analysis of the phenomenon of 4G will doubtless become available, but as yet there is nothing to hand for the interested reader by way of an interim assessment. This book sets out to fill that gap.

In essence, this book can be described as concerned primarily with the business of mobile communication and only in a subsidiary way with the technology in use. Budding mobile engineers already have access to highly technical texts, but these are largely incomprehensible to the typical business person, even if employed in the mobile industry, let alone the general public. What has not existed, prior to the publication of this book, is a review of the technology that can be readily understood by the general reader—and anyone who doubts that confusion is rife should simply bear in mind that the widely advertised 4G is in practice an inferior technology to that approved as 4G by bodies responsible for setting standards. Equally, what has not previously existed is a unified text that not merely lists all of the countries and operators that either have launched, or currently intend to launch, 4G but also examines the licensing processes involved and the financial implications.

Originally, because it was tracking events on an ongoing basis, this book developed as a continuous narrative with a single reference list. However, once the decision had been made to produce the text in book form, it became necessary to split it up into a sensible set of chapters. Unfortunately, the most obvious division of the case studies, namely into a set of regions, would have produced chapters of wildly different lengths, and hence an alternative needed to be devised.

After hiving off the discussion of technological matters into an introductory chapter, the first obvious decision was to produce the summary table of all 4G developments, listed alphabetically by country, as a second introductory chapter preceding all of the case studies. It then became apparent that the USA case study was sufficiently long to stand alone as Chap. 3. Three further case studies proved to be significantly longer than all of the others—relating to India, Russia, and the UK—and hence they were assembled together as Chap. 4. This is admittedly a slightly awkward arrangement, but it does permit readers who are interested in exploring why the introduction of 4G can become mired in controversy to obtain three contrasting views in different parts of the world.

The intervening chapters, covering Europe (Chap. 5) and the Asia-Pacific region (Chap. 6), are relatively straightforward, but Chap. 7 necessarily contains case studies covering all of the rest of the world which have been divided up into Africa, the Middle East, and the Americas. The Middle East has been treated in a somewhat flexible manner to include various CIS countries hived off from the USSR which did not sit particularly comfortably with the case studies in Chap. 5, but it is always the case that the region lying to the east of what is conventionally known as Western Europe is problematic to classify now that the European Union has encroached so far into what was formerly known as Eastern Europe. This chapter is useful in that it covers a significant number of countries that are emerging rather than fully developed economically.

We now come to the thorny issue of references. This book is based on events that are either current or very recent. As a result, there is a very limited academic/technical literature that has been published in books and journals in the public domain, and the great bulk of the information in this book has necessarily been acquired via research on the Internet. The full list of references comfortably exceeds 1,000 entries, and that does not allow for the large number of cross-references that were accessed (but are not listed) in order to verify the information contained in the references that are listed.

Clearly, to include that many references in the body of the text would both have broken up the flow of the text and used up too much space, so the somewhat arbitrary decision has been made to restrict the number of references per chapter to a maximum of roughly 100 with the exception of the relatively long chapter on Europe. This may still seem excessive to those readers who have no interest in checking the content of the websites, but the authors remain aware that an absence of references would leave them open to the accusation that they have not been either thorough or careful in determining what should or should not be included in the text. The authors have been at pains to discard those references that patently contain incorrect or ambiguous information, although some things are simply impossible to verify with certainty. The full set of references is available on the Springer website or can be obtained from the authors.

The main claim that can be made about this book is that it is entirely original in the way that a massive quantum of data has been handled. It is also unique in terms of the audience at which it is directed for whom no alternative exists. Finally, it may be asked why the decision has been made to publish at this point in time. The answer is that a sufficiently large number of what are referred to as 4G networks have now been launched to provide a satisfactory review of the state of play of what, 10 years ago, would have seemed like a futuristic era of superfast mobile connectivity. And to think that few people originally believed that mobile communication had a future of any kind.

Glasgow, UK   Peter Curwen
                                                                                      Jason Whalley

Contents

List of Tables

Technology

1

1.1 Introduction

Mobile technology is constantly evolving and it therefore comes as no surprise that it has become ever more difficult to explain the distinction between one 'generation' or 'part-generation' and another. This is aptly illustrated in the case of fourth-generation technology (4G) which is widely reported in the media as consisting primarily of Long Term Evolution (LTE). Unfortunately, this is not correct from a technical perspective as LTE should strictly be described as 3.9G. This paper sets out to provide a unique taxonomy for mobile technology in terms of 'generations' before moving on to explore the alternative versions of what is now almost universally referred to as 4G, concentrating upon the current development of LTE via numerous case studies. It continues with an exploration of the meaning of 'true 4G' and concludes with a discussion of the prospects for qualifying technologies.

1.2 Technological Issues

This paper concerns 4G but it is by no means an easy matter to explain what precisely this means. The starting point is to recognise that a mobile network can be defined in two main ways. The first concentrates upon the speed at which data are transferred, expressed in megabits per second (Mbps) while the second concentrates upon technological 'generations', commencing with 1G and working upwards. However, there is no such thing as an officially sanctioned definition of what 'generation' means in this context. There is, needless to say, a considerable overlap between these two approaches since it is possible to upgrade a technology used in an earlier generation in order to achieve speeds at least comparable to the lower range of those available via a subsequent generation. As a result, it has become necessary to reduce technology speed overlaps via the use of part-generations—for example 2.5G, 2.75G and so forth.

P. Curwen and J. Whalley, *Fourth Generation Mobile Communication*,
Management for Professionals, DOI 10.1007/978-3-319-02210-9_1,
© Springer International Publishing Switzerland 2013

Technology upgrades are achieved via improved software, hardware or both. An important point is that whereas it is quite cheap and easy to upgrade a technology (largely via software) provided it remains within the same spectrum band, it is extremely expensive to introduce a new technology in a previously unused spectrum band because a new set of hardware is required. An intermediate step in terms of cost is to open up a different spectrum band for a technology already in use, the reason being that much less new hardware will be needed.

That said, recent experience indicates strongly that operators are using the need to make their networks 4G-ready as a justification for a widespread modernisation process whereby existing 2G and 3G networks are simultaneously upgraded to provide additional capacity and better efficiency, and hence the overall cost of the project cannot be attributable specifically to the introduction of 4G.

Table 1.1 illustrates how mobile technology can be arranged within a generational framework. Several points are worthy of note. In the first place, as noted, the speed ranges of successive generations overlap to some extent. Secondly, where the same technology appears in successive part-generations, it means that it is now capable of a higher maximum speed. The most important case in point is HSPA—all technologies are written in full in the notes to Table 1.1—which can be upgraded successively from 7.2 Mbps either by doubling up the number of channels (dual-carrier) or through the use of multiple input multiple output (MIMO) antennas—see 3G.co.uk (2009b); Wikipedia (2010a)[1]; 4G Americas (2012b). Adding MIMO to HSPA helps to convert it to HSPA+, which is capable of yet higher speeds—dual-carrier HSPA+ (DC-HSPA+), for example, operates at downstream speeds of 42 Mbps or better with an uplink of at least 5.7 Mbps, and these speeds can be doubled through the introduction of 64 QAM modulation.[2]

Thirdly, as this suggests, it has been necessary in recent times to supplement the use of part-generations with the addition of '+' to denote a significant improvement in the technology, and where this appears in the table the speed overlap with the succeeding (part-) generation may be considerable.

The first line in the table denotes the progression for a standard GSM-based network as used throughout Europe and most of the rest of the world, while the second denotes the progression for a standard cdma2000-based network. cdma2000 1xEV-DO has undergone a series of upgrades which were expected to conclude with Revision C (Rev. C). In December 2006, the CDMA Development Group announced that Rev. C would be known as UMB and claimed that it would operate at up to 288 Mbps downstream in a 20 MHz bandwidth. A standard was published in September 2007 with a view to commercial availability in 2009, but as a

---

[1] Some readers may be surprised at the use of Wikipedia as a reference source given its reputation. However, this reputation has nothing to do with the quality of the technical writing which is sound in respect of telecommunications. Furthermore, Wikipedia is highly accessible.

[2] For details see Wikipedia (2010g). The technical aspects involve the bandwidth of the carrier and the use of multiple input multiple output (MIMO) technology among other matters—see also lteportal (2010). In September 2011, Nokia Siemens Networks demonstrated a 336 Mbps downlink using an eight-carrier, MIMO and 64 QAM configuration (Telecom.paper 2011c).

Table 1.1 Technological framework: the path towards LTE-Advanced

2G	2.5G	2.75G	2.75G+	3G	3.5G	3.75G	3.75G+	3.9G	4G
GSM	GPRS	EDGE	EDGE+	W-CDMA (UMTS)	HSDPA	HSPA	HSPA+	LTE	LTE-advanced
CDMA	cdma2000 1xRTT	1xRTT	–	cdma2000 1xEV-DO Rel. 0	EV-DO Rev. A	Rev. B	Rev. C	Rev. C	LTE-advanced
GSM	GPRS	EDGE	EDGE+	TD-SCDMA	HSDPA	HSPA	HSPA+	TD-LTE	LTE-advanced
PDC	i-mode	EDGE	EDGE+	W-CDMA	HSDPA	HSPA	HSPA+	LTE	LTE-advanced

Notes

cdma2000 1xRTT (radio transmission technology) is also known as IS-2000

cdma2000 1xEV-DO (evolution-data optimised) develops from Release 0 (Rel. 0) via Revisions A, B and C (Rev. A, B and C)

EDGE+ is also known as Evolved EDGE or EDGE Evolution

HSPA+ is otherwise known as Evolved HSPA, HSPA Evolution or Internet HSPA (I-HSPA)

CDMA code division multiple access, also known as IS-95 or cdmaOne, *EDGE* enhanced data [rates] for GSM evolution, *GPRS* general packet radio service, *GSM* global system for mobile, first launched by Radiolinja on 1 July 1991, *HSDPA* high-speed downlink packet access, *HSPA* high-speed packet access involving upgrades to HSDPA and use of HSUPA, *HSUPA* high-speed uplink packet access, *LTE* long term evolution, *PDC* personal digital cellular, *TD-SCDMA* time division synchronous code division multiple access, *UMB* ultra mobile broadband and is also known as EV-DO Rev. C, *UMTS* universal mobile telecommunications system, *W-CDMA* wideband code division multiple access

Source: Compiled by authors from a wide variety of sources including Wikipedia (2010a, b, c, d)

consequence of the widespread commitment by 1xEV-DO operators to move towards the adoption of LTE during 2008, UMB was effectively abandoned at the year-end.

The third line is currently restricted to China which introduced TD-SCDMA—the 'TD' refers to TDD (time division duplex) which means that the signal travels in both directions over the same bandwidth—in an attempt to prove that it was not dependent upon Western technology. Where the signal travels upstream and downstream using separate bandwidths—as, for example with W-CDMA—it is referred to as FDD (frequency division duplex). Finally, the fourth line refers exclusively to Japan where the pre-3G technology developed independently and was not adopted elsewhere. For reasons of space, other technologies in limited use are not discussed here.

Table 1.1 has been refined for this paper from previous versions published by the authors and is not as yet, therefore, in common usage, although it should be noted that no equivalent taxonomy can be found elsewhere in the public domain. Readers should be aware of the fact that both equipment vendors and network operators are inclined to hype their equipment and services—for example, all data transfer speeds are quoted by vendors as maxima, available, in practice, only if you are a single customer isolated in a laboratory. An example of the way in which operators are also deliberately muddying the ground is in respect of T-Mobile and AT&T in the USA—T-Mobile chooses to use 4G in relation to HSPA+ whereas AT&T chooses to refer to LTE-Advanced as 5G (see below).

The great majority of media reporting treats LTE and 4G as synonymous although, as shown in Table 1.1, and discussed below, this is not correct—for those with a deep interest in the technical aspects see, for example, Sesia et al. (2011); Dahlman et al. (2011). There is, therefore, a difficulty in ascribing a generational number to LTE. One approach would be to adopt the usage associated with DoCoMo in Japan, namely 3.9G (3G.co.uk 2009a), while accepting that it is not wholly satisfactory given that it is, to some extent, the result of the need to use 3.75G for HSPA+, although it must also be borne in mind that the latter is an upgradeable technology as discussed below. However, it is an unavoidable fact of life that LTE and 4G are now synonymous in the eyes of the general public, and hence there is little choice but to persevere with the term in what follows on those occasions where it is necessary to refer inclusively to LTE and equivalent technologies.

1.3   The Route to 4G

Although it is difficult to put a precise date on the emergence of 4G, it is helpful to start in March 2006 when T-Mobile, Orange, KPN and Sprint Nextel set out their joint vision in a White Paper entitled 'Next-Generation Mobile Networks. Beyond HSPA & EV-DO'. This led to the formation of the NGMN Forum in June to establish performance targets and deployment scenarios for what became LTE (3G Americas 2006). China Mobile (HK), DoCoMo and others joined the Forum with a view to creating standards for LTE by the end of 2008 and conducting operator trials during 2009. For its part, the Third Generation Partnership Project

(3GPP) has long played a very significant role in defining technical standards based upon GSM (Wikipedia 2010e, f) while its counterpart, the Third Generation Partnership Project 2 (3GPP2) serves the same purpose for systems based on cdma2000. 3GPP is usually identified via a series of Releases which were denoted by dates until 2000 and numbered consecutively starting with Release 4 in 2001 (Wikipedia 2010e). Each Release incorporates hundreds of individual standard documents which undergo a continuous state of revision. Release 7 in 2007 is primarily concerned with HSPA+ (3G Americas 2007) while those commencing with Release 8 are concerned with the route to 4G.

The timetable set out in Release 8 (2008Q4) onwards placed LTE roughly 1 year behind WiMAX (discussed below) in terms of development. DoCoMo was keen to pre-empt the discussion, and to this end issued its proposed pre-standard version of LTE called Super 3G (3.9G in our taxonomy). However, other members of the Forum were not keen on this arrangement. As a result, a LTE/SAE Trial Initiative was set up in May 2007 to demonstrate the potential of LTE and SAE (system architecture evolution) through joint tests, including radio transmission performance, early inter-operability, field trials and full customer trials (Global mobile Suppliers Association 2010a). The Initiative was expected to last for up to 2 years. As part of this, Alcatel-Lucent and LG Electronics announced the completion of test calls in November 2007 while Nokia Siemens Networks (NSN) completed the first multi-user field trial in December, following up with the first call using a commercial base station and fully compliant (with 3GPP Release 8) software in September 2009.

Finland was the first European country specifically to allocate new spectrum for 4G to incumbents in the GSM band without the need for new licences (Global Insight 2009)—it is worth noting that whereas LTE supports channel bandwidths from 1.4 MHz to 20 MHz, a wide channel of between 10 MHz and 20 MHz is preferable because it allows for higher speeds—although it also assigned licensed spectrum in the 2,500–2,690 MHz (2.6 GHz) band in November 2009 (see below). LTE was expected primarily to utilise existing GSM spectrum together with the 2.6 GHz band although, in the USA, Verizon Wireless (and to a lesser extent AT&T) intended to use the 700 MHz spectrum won in the course of Auction 73 (Federal Communications Commission 2007).

The dominant role currently being played by the 2.6 GHz band reflects the fact that it is being used throughout the European Union (EU) in accordance with ECC Decision (05)05 of 18 March 2005, 'On harmonized utilization of spectrum for IMT-2000/UMTS systems operating within the band 2,500–2,690 MHz'. The EU, unlike the USA, had also mandated common spectrum for all 3G launches, and its action in respect of 4G spectrum reinforces the point that although one regional group cannot enforce common spectrum use upon other such groups, if a sufficient number of large countries are early adopters of a spectrum band then other countries are under pressure to follow suit because it facilitates international roaming.

1.4   TD-LTE

It is of interest that, despite being forced to deploy TD-SCDMA as its 3G technology when the Chinese mobile industry was restructured, China Mobile also committed to launch LTE in the guise of TD-LTE in 2011. In March 2010, Clearwire in the USA—ostensibly the main exponent of WiMAX there (see below)—added its voice to those of the likes of China Mobile, Huawei, ZTE and Motorola in asking 3GPP to define 2,500–2,690 MHz as a TDD band for LTE. For the USA specifically, the request was for the entire band to be opened up for TD-LTE as against the 2,570–2,620 MHz band (Band 38) already authorised (combined with 2,500–2,570 MHz for FDD uplink and 2,620–2,690 for FDD downlink, thereby minimising the number of FDD/TDD borders which tend to need buffer channels to avoid interference and maximising the efficient use of the spectrum). However, the largest suitable contiguous bandwidth (Band 40 comprising 100 MHz) lies within the 2.3 GHz band which can be divided efficiently into five 20 MHz channels. The fact that developments are taking place within Bands 38 and 40 is a significant advantage compared to FDD versions of LTE which are taking place within a multiplicity of bands.

According to Global mobile Suppliers Association (2010b), testing of TD-LTE had already taken place, often in conjunction with tests of LTE using paired spectrum, in China, France, India, Ireland, Japan, Malaysia, Oman, Poland, Russia, Taiwan and the USA—see also lteportal (2011: p. 43). In Denmark and Sweden, Hi3G declared its intention to launch using both FDD and TDD in the 2.6 GHz band. However, it is worth noting that the Chinese government considers TD-LTE to be an immature technology and, in conjunction with the need to obtain a return on the relatively recent investment in 3G, has determined that no TD-LTE launch will take place before 2014 (Dow Jones 2011).

A recent development has been the formation of the Global TD-LTE Initiative (GTI) in February 2011 (Technology Marketing Corporation 2011). The founding members were Aero2, Bharti Airtel, China Mobile, Clearwire, E-Plus, Softbank Mobile and Vodafone, and a major thrust was to achieve convergence with LTE FDD in order to maximise economies of scale as well as a sharing of the ecosystem with other TDD technologies such as the Japanese eXtended Global Platform (XGP). As part of this initiative, China Mobile and Clearwire, together with other GTI members, agreed on common test specifications and joint interoperability testing, particularly in the 2.3 GHz to 2.7 GHz bands (Telecom.paper 2012a). Separately, China and Taiwan signed an MOU in September 2012 with a view to collaborative research (Gabriel 2012f).

There is some progress in the less popular 3.5 GHz band where ZTE has unveiled the world's first high-power base station and signed contracts to roll out networks in the Asia-Pacific region and Africa (Gabriel 2012c)—for a list of actual and proposed launches in all relevant bands see Global mobile Suppliers Association (2012d).

Overall, there is some dispute about the prospects for TD-LTE, but with increasing numbers of operators indicating that they intend to proceed on a multi-mode basis—that is, FDD combined with TDD—Ovum, for one, is predicting that TD-LTE will provide one-quarter of all LTE connections by 2016 (Middleton 2012). The first bi-directional handover between paired and unpaired spectrum took place on the China Mobile network in Hong Kong in June 2012 [TeleGeography 2012b and see also Cellular-news (2013b)]. However, until such time as suitable smartphones reach the market there will be only limited interest from consumers.

As of November 2012, when the Global mobile Suppliers Association published its latest update—see also the case studies in the chapters that follow—11 TD-LTE networks had been launched in ten countries. However, these launches involved five different 3GPP spectrum bands and some fairly peripheral players. Even if those companies that have declared an intention to launch in the future are included, the number of 2G/3G incumbents is very much in the minority. It would accordingly be unwise to expect dramatic developments any time soon.

Meanwhile, signs are emerging that versions of TD-LTE will become available that meet the technical requirements of LTE-Advanced. For example, in September 2012, Nokia Siemens Networks claimed that in laboratory conditions it had achieved a 1.6 Gbps data transfer using a 60 MHz carrier. Specifically, it used an 8-pipe radio module enabling eight streams of uplink MU-MIMO (Wikipedia 2012a) compared to the four streams specified for LTE-Advanced. Subsequently, China Mobile combined with Ericsson to demonstrate a 223 Mbps downlink using two 20 MHz carriers (TeleGeography 2012d) whereas in June 2013 Nokia Siemens Networks achieved a record 56 Mbps uplink using Single User MIMO (SU-MIMO) and a single 20 MHz carrier (Sahota 2013). It is also worth noting that it was China Mobile's pre-commercial trial network in Xiamen that was used to demonstrate the world's first live TV broadcast via TD-LTE.

Whereas the WiMAX Forum was initially disposed to see TD-LTE as a rival to fight off, it is now resigned to the fact that the battle is already largely lost and hence that its best interests lie in integrating TD-LTE into the specifications for WiMAX2—see below and Gabriel (2012g).

1.5   LTE in the 1,800 MHz Band

The growing importance of the 1,800 MHz band partly reflects delays in licensing bands set aside primarily for LTE (and/or WiMAX)—most obviously the 2.6 GHz band, which also requires many more base stations to achieve the same coverage—and also delays in the availability of spectrum in the previously favoured digital dividend band—that is, spectrum released by switching off analogue radio signals (4G Americas 2011). Huawei, among others, was quick to foster this development through its FlexiRAN software which supported, *inter alia*, seamless handover between HSPA at 2.1 GHz (used for 3G) and LTE at 1,800 MHz. Finland was the first country to permit the use of the 1,800 MHz band in 2009, since which time there have been numerous other launches—see Table 1.2 and Global mobile Suppliers Association (2012b).

Aside from the coverage issue, the 1,800 MHz band has rapidly become favoured for LTE because it is used for 2G throughout Europe and in sufficient other parts of the world to foster international roaming. Furthermore, most operators have spare capacity in the band—they often rely mainly on the 900 MHz band for 2G.

Table 1.2 3GPP Bands for TD-LTE	3GPP Band	Frequencies (MHz)[a,b]	Region
	33	1,900–1,920	Europe, Asia (not Japan)
	34	2,010–2,025	Europe, Asia
	37	1,910–1,920	Former PCS-band (US)
	38	2,570–2,620	Europe
	39	1,880–1,920	China
	40	2,300–2,400	Europe, Asia
	41	3,400–3,600	–
	42	3,600–3,800	–

[a]As with paired (FDD) spectrum, TD-LTE is designed to operate using six carriers comprising 1.4, 3, 5, 10, 15 and 20 MHz of bandwidth. Larger bandwidths enable higher data transfer rates
[b]With unpaired spectrum, the uplink and downlink can be asymmetrical with the downlink up to nine times larger than the uplink
Source: Adapted from Hetting and Stanislawski (2010)

In April 2011, the European Commission approved, by way of a Decision that extended the remit of the GSM Directive, technical rules on how the 900 MHz and 1,800 MHz bands should be opened up for LTE and WiMAX. National regulators were given until 31 December 2011 to implement the Decision (Telecom.paper 2011a). However, the 900 MHz band holds little appeal, given the paucity of spare capacity to re-farm and the proximity of the digital dividend band in Europe.

1.6   LTE in the Digital Dividend Bands

Certain spectrum bands are popularly referred to as 'digital dividend' because they are bands where spectrum can be released by switching from analogue to much more efficient digital signals. The analogue signals are predominantly used for broadcast TV, and it is the ambition of all advanced (and many emerging) countries to switch over to wholly-digital broadcasting by the mid-decade. However, the process is not going to take place in one fell swoop, nor is there necessarily going to be consistent use of spectrum within a given region or even country, because the spectrum involved spans a wide range of (possibly overlapping) frequencies. For example, the complete 700 MHz band is 698–806 MHz in the USA, but this is divided into several sub-bands and, in effect, the main incumbents have sought to annex different sub-bands thereby creating difficulties in roaming across networks (see Chap. 3). Meanwhile, 800 MHz translates as the 790–862 MHz band in Africa, Europe and the Middle East. In contrast, the digital dividend band in most of Asia spans 470–960 MHz although many individual countries (most recently Brunei, Indonesia, Malaysia and Singapore) have opted for the 700 MHz band, specified by the Asia-Pacific Telecommunity (APT) as 703–748 MHz paired with 758–803 MHz—see Global mobile Suppliers Association (2010b); 4G Americas (2011: pp. 64–9). The Mexico case study in Chap. 7 is illustrative of the growing enthusiasm for the APT plan in Latin America where there is little or no desire to copy the USA plan.

As per usual, the European Union seeks to impose harmonised use of digital dividend spectrum across all member states. The European Commission, together with the Council and Parliament, agreed a text which required national regulators to authorise the use of digital dividend spectrum by January 2013. This formed part of the Radio Spectrum Policy Programme which set out the target of making at least 1,200 MHz of spectrum available for high-speed broadband by 2015. Naturally, it is one thing to authorise the use of a band and another to clear it of existing users, a problem that is sensitive in this case because, firstly, there may be military usage and, secondly, it means switching off the analogue signal for TV and forcing consumers to purchase a new digital television which they cannot necessarily afford. While it is not particularly costly to use subsidies in certain deserving cases, it is evident that a reasonably long warning period must be provided and so far digital dividend spectrum does not play a role in Table 1.2.

Nevertheless, this spectrum will sooner or later play a major role in the provision of 4G. Some interesting case studies by Aetha on behalf of the Global mobile Suppliers Association (Aetha 2011a, b, c, d, e) indicate why licensing this spectrum is problematic, but at the end of the day the need to open up new spectrally efficient bands with sufficiently wide coverage to permit roaming is an imperative if data are to be transferred at even remotely near maximum 4G speeds.

As of end-2012, a sufficient number of 800 MHz spectrum auctions had taken place in Europe to throw up some interesting insights (Aetha 2012). In particular, it is noteworthy that by far the most common outcome was for there to be three winning bids even though there were generally four 2G incumbents, and that in half the auctions the fourth-largest operator declined even to register a bid—which was probably a sensible strategy as it failed to win a licence on every occasion that a bid was registered despite having a shortage of spectrum at the time.

However, progress is not all it should be and the European Commission was obliged to allow nine Member States to delay well beyond the original January 2013 deadline for auctioning spectrum in the band with a further five registering applications to that end (Hibberd 2013).

Use of the 800 MHz band is more problematic in the case of Africa because there is a relatively wide provision of fixed-wireless in the band and it is not intended that this should be replaced wholesale with fully mobile provision. Hence, only the 790–806 MHz band is strictly available as a digital dividend. The situation in the Middle East is not much better.

This partly explains why attention has recently switched to the potential of the 700 MHz band which spans 694–790 MHz in ITU Region 1—Europe, the Middle East, Africa and Northern Asia. This band was expected to come up for discussion at the World Radiocommunication Conference (WRC) in 2015, but the preliminary groundwork has been brought forward in the hope that all regions will be able to agree on harmonised use of the band at WRC-15. At WRC-12, the decision was taken to co-allocate it by 2015 for mobile and broadcasting services, but it was not established whether it would be used for FDD or simply as a supplemental downlink.

The European Commission has been turning its attention to the 700 MHz band via the Radio Spectrum Policy Group (RSPG). The Digital Agenda Commissioner of the EU is very keen to harmonise the band across Europe while giving all interested parties a fair allocation. Not surprisingly, European broadcasters using the band are less than enthusiastic about encroachment by mobile operators, but the significant overlap with the APT plan—the favoured plan is to combine a 703–733 MHz uplink with a 758–788 MHz downlink—is a major incentive with a view to achieving economies of scale in equipment provision. It may be noted that the USA plan is being studiously avoided in order to prevent incumbents there from dictating equipment availability. However, given that many European countries are grappling with problems in developing the 800 MHz band, it is unlikely that they will be rushing to open up the 700 MHz band as well.

One country that appears to want to gain a first-mover advantage is Russia, where 7.5 MHz paired in the 700 MHz band was auctioned off in July 2012 (see Chap. 6). The underlying belief appears to be that only relatively low frequencies can be used in rural areas and for high-quality indoor coverage in an environment which is increasingly hostile to the construction of new base stations.

1.7   LTE in the 450 MHz Band

Although the 450 MHz band is being used for 3G, this generally involves somewhat peripheral operators and is largely restricted to rural coverage. In good part, this reflects the reliance upon CDMA450 technology which, in turn, reflects the reluctance of vendors to invest in providing W-CDMA/UMTS equipment compatible with this spectrum. The region which provides the best opportunity to achieve a scale of operations that might prove attractive to a vendor is the former Eastern Europe, since CDMA450 was sometimes introduced there before the desirability of GSM-based networks, driven by the need to roam, became paramount.

Accordingly, in September 2010, five East European operators with existing CDMA450 networks—Dialog (Belarus), SkyLink (Russia), TeleMobil (Romania), Triatel (Latvia) and Ufon (Czech Republic), together with G-Mobile (Mongolia), approached their national regulators and the ITU with a view to obtaining permission to re-farm their networks for LTE. Interest has also been expressed by MegaFon and VimpelCom in Russia where, as noted below, LTE licensing is in a state of flux (Gabriel 2010b).

However, the 450 MHz band was also in common usage for the analogue technology NMT which has finally been shut down in its entirety—the last one was closed in Poland in June 2012 giving Orange a splendid opportunity to switch to CDMA450 which is roughly 12 times as efficient in terms of geographic coverage as equivalent technology using the 1,800 MHz or 2.1 GHz bands. This, in turn, helps explain why 3G has largely been exploited in densely populated urban areas, leaving more rural areas to be covered with EDGE by GSM-based operators. Given the need for far fewer base stations, CDMA450 has a built-in cost advantage that can be exploited when LTE is under consideration.

At the end of the day, if an operator cannot make an economic case for rolling out HSPA+ in rural areas using spectrum bands of 800 MHz or more, it is unlikely to want to use LTE either unless it is put under pressure by a regulator to do so prior to being authorised to provide LTE in urban areas (as in Germany). This suggests that the 450 MHz band has a reasonably bright future for LTE use, especially since it has not even been licensed for cellular use in many countries.

The main problem is that LTE in the 450 MHz band has be authorised by the relevant standards bodies which is not imminent. Without that authorisation, vendors lack the incentive to develop multi-band devices that include the 450 MHz band. It is slightly ironical in this context that world leaders in LTE in the likes of South Korea and the USA started out with CDMA technology. What the future holds for LTE in the 450 MHz band is for now a matter of speculation, but significant developments are expected within the next several years (Middleton 2013).

1.8 Reassigning Spectrum

In the course of the case studies below, reference is made in passing to the extension of existing licences and to the re-assigning of licences that have reached the end of their term. This is distinct from a situation where the original use for the spectrum has been terminated and it is redeveloped for a different purpose via a 'primary assignment', usually in the form of an auction. Clearly, what may usefully be termed a 'primary re-assignment' (Feasey 2012) is likely to play a less significant role in the overall pattern of spectrum use for 4G, but it is worth making a few comments at this point because, coincidentally, many of the original 2G licences are about to expire.

It is worth noting that almost nothing was written into the original contracts for 2G licences about what would happen when they expired, presumably because it was assumed that they would simply be rolled over (as was explicitly made possible in certain cases). Where spectrum rights are simply renewed the term 'grandfathering' is used, but there is nothing in principle to prevent the regulator from re-assigning by administrative fiat or re-selling expiring licences (Vesterdorf 2012). At the same time, the regulator may try to alter the competitive landscape by imposing spectrum caps, taking into account the way in which the spectrum has been used (or not) during the original licence period.

When detailed discussion about spectrum use for 4G became necessary, most regulators were inclined to assume that, as with 3G in Europe, new spectrum bands would be opened up, either in the lower bands such as 700 MHz or higher bands such as 2.6 GHz. In practice, however, it rapidly became clear that because these bands would have to be cleared and subsequently sold, this could turn out to be a long-winded process—see, for example, the UK case study in Chap. 4. Hence, operators exerted pressure for existing bands—and in particular the 1,800 MHz band—to be re-farmed for LTE. For the most part this was uncontroversial, but it obviously created a difficulty in that many of the relevant licences were near, or at, their expiry date. Hence, simply re-assigning them at the going rate to existing licence holders would provide them with a competitive advantage. As the case studies demonstrate, different countries have adopted a variety of approaches to resolving this issue.

1.9   Device Certification

Operators have the choice when launching a network whether to use devices directed primarily towards the corporate market, such as data cards and dongles, or to introduce handsets directed primarily towards the consumer market or to do both simultaneously. This obviously depends in part on what devices are available, which in turn depends upon whether devices have satisfied the Certification Criteria of the Global Certification Forum (GCF)—for details see http://www.globalcertificationforum.org. As of July 2011, the GCF had sanctioned FDD devices for the 2,100 MHz band (Band 1), the 2,600 MHz band, (Band 7), the upper 700 MHz North American band (Band 13) and the 800 MHz European Digital Dividend band (Band 20) (Cellular-news 2011a).

1.10   WiMAX

Another way forward for 4G is encapsulated in the Worldwide Interoperability for Microwave Access (WiMAX) Forum which is behind the Institute of Electrical and Electronics Engineers (IEEE) 802.16-2004 *fixed* standard first published in April 2002 (WiMAX Forum 2009). This is a data-only standard—not strictly a technology as such—that primarily covers the 2 GHz to 11 GHz band, making it suitable for connection to Wi-Fi hotspots, and the 56 GHz to 66 GHz band. The *mobile* variant of WiMAX in current use is known as 802.16e-2005 and was approved in December 2005.

With a peak range of at least 50 km—and hence suitable for metropolitan area networks (MANs)—and a peak shared data rate of 70 Mbps, 802.16e certainly appeared to offer an attractive prospect for areas currently either not covered at all, or poorly covered, by other types of network. However, this speed reflects the optimum channel size, power and best-case modulation, and takes no account of the network overheads (Wikipedia 2010d). In fact, roughly 30 % of the bandwidth is needed for error correction and encoding, and what is left—roughly 50 Mbps—covers both directions of data transfer given that WiMAX is a TDD system. Hence, with equal directional flows, each could carry only 25 Mbps, and a further 10 % is needed for other purposes. What is left—20 Mbps downstream and upstream—is accordingly rather less than the original amount claimed, and given that WiMAX is a point-to-multipoint network, the result of sharing the available spectrum among all those logged on at any one time means that a realistic downstream speed in mid-2010 was 3–5 Mbps combined with 1 Mbps upstream.

However, as noted, no technology delivers maximum cited speeds so the more important issue is whether WiMAX can be considered as a 4G technology. It may be noted that, in June 2007, the International Telecommunication Union (ITU) gave its initial approval to plans to include a specific subset of WiMAX known as orthogonal frequency division multiple access time division duplexing (OFDMA TDD WMAN) as a terrestrial radio interface under IMT-2000 (which governed 3G), meaning that spectrum would be reserved for this on an international basis (Global Insight 2007a). Final approval was forthcoming in October with the result that WiMAX could henceforth be rolled out in the 2.6 GHz band and the 1.9 GHz 3G band. But in gaining acceptance as a 3G technology, WiMAX appeared to have given up its pretensions to be treated as part of the 4G family of technologies despite the high theoretical speeds of which it is capable (Global Insight 2007b). For its part, the WiMAX Forum has sanctioned licensing in the 2.3 GHz, 2.5 GHz and 3.5 GHz bands. Unfortunately, there is no agreement on a single spectrum band to be used on a world-wide basis—the USA favours 2.5 GHz, Asia-Pacific countries favour 2.3 GHz and 2.5 GHz and Europe and Latin America favour 3.4–3.6 GHz.

1.11   VoLTE and Rich Communication Services

Given that LTE is a data-only technology, among the outstanding technical issues to be fully resolved is how to run voice services over LTE (VoLTE)—for the initial discussion see, for example, Poikselkä et al. (2012); GSM World (2010); Cellular-news (2011b)—and how to improve the signal-to-noise ratio. 2G and 3G were designed primarily to carry voice rather than data, but 4G is the reverse and based upon Internet Protocol standards. At present, most LTE networks where the operator also provides HSPA use the interim Circuit Switched Fallback solution whereby LTE is used for data and the signal is switched to HSPA for voice—although for cdma2000 networks both radio interfaces must be used simultaneously which is a heavy drain on handset batteries (Khan 2012)—a problem that is being addressed with some success (Cellular-news 2013d).

A big advantage of VoLTE is its ability to handle high-definition voice (HD Voice) and to provide operators with an in-built alternative to 'over-the-top' services such as Skype. HD Voice is the term commonly used in Europe for a version of wideband audio. This has its origins in a decision by the ITU to standardise a version of wideband audio as G.722 in 1987. In effect this uses a speech compression algorithm to extend the range of audio signals from a narrowband range of 300 Hz to 3.4 kHz, far narrower than the potential range of the human voice, to a range of 50 Hz to 7 kHz. Nokia and VoiceAge developed a version of G.722 known either as Adaptive Multi-Rate-Wideband, Adaptive Multi Rate Wideband or Advanced Multirate Wideband (AMR-WB or W-AMR)—the use of differing versions for an agreed set of initials or even differing initials is not uncommon in telecommunications—which was subsequently taken up by 3GPP (Wikipedia 2012b). VoLTE is defined in GSMA specification VoLTE IR.92 which is based on 3GPP standards.

HD Voice is being introduced fairly rapidly. According to the GSA in September 2012, from an initial launch in 2009 by Orange in Moldova, the number of service launches had reached 45 in 35 countries, primarily in Europe (Global Mobile suppliers Association 2012a). Orange is the leader in the field, with service launches in double figures, and it is trying to get other European operators to follow suit.

It must be borne in mind that HD Voice can be introduced in conjunction with any mobile technology, not simply LTE. The first live VoLTE call over a commercial network was allegedly placed by Verizon Wireless in February 2011 (Gabriel 2011b) although this is in dispute (Weaver 2011) and Verizon Wireless does not anticipate a commercial launch until 2013. In contrast, Huawei had already produced a triple-mode LTE modem which, by connecting GSM, UMTS and LTE, allows a customer to access LTE downlink speeds while a network is still being rolled out (Cellular-news 2010). The first commercial VoLTE launches were by SK Telecom (using the Samsung Galaxy S III) and MetroPCS (using the LG Connect 4G Android) in early August 2012 (TeleGeography 2012c; 4G-Portal 2013b), closely followed by LG Uplus (in Seoul, using the LG Optimus LTE2). The initial launches and deployments were largely restricted to North America and South-east Asia (Global mobile Suppliers Association 2012a).

In December 2012, Sweden's Tele2 claimed that it was the first operator in Europe to test VoLTE and expressed the hope that it would be able to launch before the end of 2013 (TeleGeography 2012e). For its part, Telefónica in Germany used a test laboratory environment to demonstrate (a world first) the SRVCC (Single Radio Voice Call Continuity) standard which allows for the handover of a VoLTE call from a LTE network to 3G (Cellular-news 2013a). VoLTE is expected to generate revenues in the billions within a few years (Cellular-news 2012b).

It is now commonplace to discuss VoLTE in conjunction with Rich Communication Services (RCS) which are marketed by the GSM Association as 'Joyn'. The idea is to combine VoLTE and RCS in order to create a full rich messaging and multimedia communications service spanning 3G and 4G. The main features of RCS are an enhanced phonebook, enhanced messaging (including messaging history) and an enriched call (multimedia content sharing during a voice call).

The RCS initiative, originally called RC Suite, began in 2007 and was taken under the auspices of the GSM Association in 2008. In 2011, RC Services became the preferred title for a new specification known as RCS-e (RCS-enhanced) (Wikipedia 2013a). So far, there have been five Releases for RCS, with Release 4 dealing with support for 4G. In June 2012, Orange, Telefónica and Vodafone in Spain were formally accredited as RCS-e (Joyn) networks—the first Joyn-embedded handsets were launched in April 2013 (Telecom.paper 2013a)—with operators in Germany, France and Italy also committing to Joyn. In Asia, SK Telecom, Korea Telecom and LG Uplus have set up RCS and are working on inter-operability. Unlike the other operators which intend to provide a free service in order to wean customers from the likes of Skype, SKT intends to levy charges for certain features of Joyn. In the Middle East, Kuwait's Zain was the first to commit to Joyn via its Partner Market agreement with Vodafone.

1.12 'True' 4G/IMT-Advanced

As indicated above, LTE has been described as 4G for marketing purposes but in reality falls short of the technical specifications in important respects (Wikipedia 2010g). In practice, there are two main contenders for the right to be called 'true' 4G, namely LTE-Advanced (LTE-A) and WiMAX2. The term IMT-Advanced is often used as a synonym for 'true' 4G.

Hence, despite the comments above, WiMAX is now being pushed heavily as a 4G technology. The WiMAX2 Collaboration Initiative was set up in April 2010. It consisted of ten major companies involved in 802.16 including Intel, Samsung, Motorola, Alvarion and ZTE. Its purpose was essentially to support the move up from 802.16e to 802.16m, otherwise known as WiMAX2, via a collaboration with the WiMAX Forum which announced that it expected WiMAX2 to be completed before the year-end with device certification coming in late 2011 (Gabriel 2010a). 802.16m is capable, in theory, of a 300 Mbps downlink (Gabriel 2010c)—a comparison with LTE-Advanced is to be found in Dailywireless (2010). The term WirelessMAN-Advanced is often used as a synonym for WiMAX2 and is favoured by the ITU.

As noted, both 3GPP and the ITU are keen to have a say in what is meant by 'true' 4G. 3GPP, which covers the development only of GSM-based technology, released the specifications for 'true' 4G within Release 10 and Beyond. In essence, for the 3GPP to sanction it, 'true' 4G must deliver a 100 Mbps downlink with high mobility and wide area coverage, a 1 Gbps downlink and a 500 Mbps uplink when stationary, low latency of under 10 ms round-trip delay and use wide spectrum bands of up to 100 MHz (Global mobile Suppliers Association (2012c) and see http://www.radio-electronics.com). Such a wide band can only be achieved via LTE Carrier Aggregation—that is, by combining carriers/channels in order to build up the bandwidth along which signals can be carried (Arora 2013). Release 10 specified 'component carriers' of up to 20 MHz so the optimum manner to achieve a 100 MHz bandwidth is to aggregate five 20 MHz carriers (4G Americas 2012a). The above requirements are not met by LTE (which meets the specifications of Release 8 and Release 9) but they are met by LTE-Advanced.[3]

For its part, the ITU operates through its Radiocommunications sector (ITU-R) which, in late October 2010, chose LTE-Advanced and WiMAX2 as its official candidates for 'true' 4G (Gabriel 2010d). Final ratification was pencilled in for the

---

[3] For those readers of a technical disposition, an excellent introduction to LTE-Advanced is to be found in Radio-Electronics.com (2010). A wide variety of technical matters relating to LTE are available at http://www.lteworld.org. Carrier Aggregation can take three forms: contiguous intra-band, non-contiguous intra-band and inter-band. Release 10 contains two intra-band sets (2.1 GHz and 2.3 GHz) plus one inter-band set (2.1 GHz/800 MHz). Work on adding further sets is proceeding at 3GPP but there are those who are somewhat sceptical about the process, noting that vendors will struggle to satisfy many of the potential combinations and that small cells can provide an alternative means for satisfying consumers' expectations.

ITU-R Study Group 5 meeting in late November.[4] However, the list of potential qualifying technologies was not pre-judged, and in practice six were proposed including WirelessMAN-Advanced and LTE-Advanced using both FDD and TDD (Morris 2010) for consideration in early 2011. WiMAX2 was the first to be ratified in early April (Gabriel 2011a). In January 2012, the agreed technical specifications for IMT-Advanced were finally ratified by the ITU Radiocommunications Assembly (for details go to http://www.itu.int), and LTE-Advanced and WirelessMAN-Advanced were accepted as meeting those specifications (Telecom.paper 2012b).

As usual, Japan was in the forefront of developments. In late January 2011, the Ministry of Internal Affairs and Communications issued a licence to DoCoMo to commence pilot trials of LTE-A, the company having by this time demonstrated a 1 Gbps downlink and a 200 Mbps uplink in a laboratory setting (Global mobile Suppliers Association 2012c). A full trial licence was expected to be granted in March. Predictably, South Korea was not far behind with governmental agencies demonstrating a network capable of a 400 Mbps downlink in February 2011. Two Asian vendors, ZTE and Huawei, both claimed to have been the first to have implemented inter-band Carrier Aggregation combined with MIMO in February 2012, thereby permitting a theoretical maximum downstream speed of 270 Mbps, in the case of ZTE using 20 MHz channels in the 1,800 MHz and 2.6 GHz bands and in the case of Huawei using the 800 MHz and 2.6 GHz bands (Gabriel 2012a). In Europe, also predictably, Ericsson of Sweden was at the forefront of developments, demonstrating LTE-A using a 3 × 20 MHz aggregated carrier and 8 × 8 MIMO in June 2011 (Telecom.paper 2011b). Somewhat confusingly, ZTE and China Mobile claimed to have carried out the world's first commercial Carrier Aggregation test at the end of January 2013 within an outdoor TD-LTE environment achieving a single-user downlink of 223 Mbps (Cellular-news 2013a). This has been bettered more recently—for example, during a field trial by Huawei using the Touch network in Lebanon in April 2013 (Telecom.paper 2013b).

To conclude this section, a salutary reminder that technology waits for no man is in order. Although the world has largely come to terms with LTE and is just about getting to grips with LTE-A—an example of progress was the demonstration of a core LTE-A technology known as Transmission Mode 9 (TM-9) by Ericsson and SK Telecom in January 2013 (4G-Portal 2013a)—it really is not ready for yet further advances in technology. This is unfortunate because, in March 2012, Huawei introduced 'Beyond LTE', its self-explanatory term for its latest batch of patents underpinning a prototype network capable of reaching a peak data download rate of 30 Gbps (Gabriel 2012b). It followed up in May by announcing successful tests over the 2.6 GHz band LTE networks of Tele2 and Telenor in Sweden [acting jointly as Net4Mobility see TeleGeography (2012a)] which had achieved a downlink of 290 Mbps. It also announced that it had conducted a field test, using an unspecified commercial European network, which for the first time involved a Category 4 device

---

[4] Information on a number of relevant themes can be found on the ITU website, specifically in this case at http://www.itu.int/ITU-R

(one capable of a downlink of at least 150 Mbps). This incorporated the first HiSilicon LTE multi-mode chipset to support Release 9 (Gabriel 2012d).

As well as consideration of what the future holds, it is perhaps useful to note where precisely things stand as of July 2013. The first Category 4 devices were launched towards the end of 2012. However, the first one of these in the form of a handset, the Samsung Galaxy S4 LTE-A, was used in June 2013 when SK Telecom launched the first LTE-A service in central Seoul and 42 other cities providing a maximum downlink of 150 Mbps, twice that being provided previously. The technology combined Carrier Aggregation with Coordinated Multi Point (CoMP) software, alternatively known as co-operative MIMO [see, for example, Irmer et al. (2011)], and is to be upgraded with Enhanced Inter-Cell Interference Coordination (eICIC) [see, for example, Kimura and Seki (2012)] in 2014. It should be noted that this is in essence a process of adding technological advancements as part of a rolling programme. The switch to LTE from HSPA was a step-change in technology, but the upgrade from LTE to LTE-A does not follow the same pattern. There is no obligation on an operator to adopt a particular pathway, but as indicated almost (if not) all operators are expected to use Carrier Aggregation as the first step (Gabriel 2013c; TeleGeography 2013).

The next development beyond LTE-A (covered by LTE Release 10 and Release 11) is commonly referred to as LTE-B and encompasses what is to be contained in LTE Release 12 and Beyond on which work has begun in 3GPP—see http://www.3gpp.org/Release-12. As of January 2013, work on Release 11 has yet to be completed but mostly Release 11 performs the role of tweaking what is contained in release 10. Release 12 is discussed in Ericsson (2013) which is in general too technical to warrant detailed analysis at this point. It may be noted, however, that the initial sections comprise 'Further enhanced multi-antenna transmission', 'Further advanced terminal receivers' and 'New carrier type'. The paper goes on to examine 'Small-cell and local area deployment' where it is noted that Release 12 seeks to achieve an even higher degree of inter-working between the macro and low-power layers (such as 'hotspots') including dual-layer connectivity. In essence, what this means is that a device connects simultaneously to both layers with, for example, the downlink arriving via one layer while the uplink is sent via the other layer or control signalling being provided by the macro layer while connectivity is achieved by the low-power layer. This, in turn, raises issues about the role to be played by Wi-Fi which currently provides a signal in certain locations provided by an array of (often retail) sources. However, a device will not necessarily pick up a stronger signal than can be provided by a LTE network when it locks onto a Wi-Fi signal.

In February 2013, the European Commission provided €50 million to fund research with a view to delivering so-called 5G technology by 2020 (Cellular-news 2013c). Subsequently, Broadcom and SK Telecom claimed to have launched a 5G Wi-Fi router (Telecom.paper 2013c), but there must be a suspicion that if the term '4G' has been open to abuse, this will be even more the case with '5G', particularly during the period prior to specifications being agreed by 3GPP and the ITU. One problem, as ever, is that various vendors are cracking on regardless. Of particular interest is Samsung's experimental system, announced in May 2013 (Gabriel 2013b) which concentrates upon high-frequency spectrum. This is normally

used for access purposes—that is backhaul or fixed-wire connectivity—but Samsung
has managed to achieve a 1 Gbps transmission using the 28 GHz band using adaptive
array transceiver technology. In essence, the technology uses 64 antenna elements to
reduce the radio propagation loss at millimetre-wave bands, thereby improving both
range and indoor penetration. However, this technology will not be commercially
available until roughly 2020.

1.13   Long Term HSPA+ Evolution

At this point, it is worth reminding ourselves that it is possible, in principle, for a
technology which fails to meet the requirements of 'true' 4G to operate at equivalent speeds. As noted, LTE-Advanced was required to sustain a downlink of
100 Mbps, but during 2010 vendor Huawei claimed that it would be able to upgrade
HSPA+ to provide a downlink of 300 Mbps. Not to be outdone, Nokia Siemens,
acting in conjunction with operator T-Mobile USA, floated the possibility in
December 2010 of a downlink of 672 Mbps, to be achieved by combining up to
eight carriers and improving spectrum utilisation. This was described as Long Term
HSPA+ Evolution, and it was suggested that it could come on stream commercially
in 2013 or 2014 in parallel with LTE-Advanced. It also happened to provide T-
Mobile USA with an excuse to describe its HSPA+ network as '4G' even though,
at the time, the network was operating at nowhere near 'true' 4G speed.
    The 3GPP time sequence for HSPA downlinks is broadly as follows: Release
7 = 21/28 Mbps; Release 8 (dual carrier) = 42 Mbps; Release 9 (dual carrier) = 84
Mbps; Release 10 (quadruple carrier) = 168 Mbps; forthcoming Release 11 (octuple
carrier) = 336 Mbps (Cellular-news 2011c). The uplink for Releases 9 and 10 is
23 Mbps and for Release 10 is 46 Mbps. The position as of end-2012 is that Release 11
has been functionally frozen since September, meaning that no additional functionality
can be added to the Release. However, agreement on all technical specifications was
likely to take until March 2013.

1.14   Small Cells

The huge demands placed upon networks in densely populated urban areas have
generated a need for solutions involving small cells. Clearly, Wi-Fi is already in
quite widespread use as a means for downloading data, but it is often provided by a
variety of suppliers with individually patchy coverage. It is accordingly likely that
new solutions will emerge which may well integrate Wi-Fi within the overall
provision based upon dense networks of tiny base stations—as in the case of
AT&T (Gabriel 2013a).
    The integration of Wi-Fi with small cells (femtocells) is likely to commence
with a common core and progress to the base station level. Many vendors are
involved such as Nokia Siemens (NSN) which is launching Smart Wi-Fi, building
on its existing FlexiZone architecture (White 2013a). This is a highly technical

issue and hence cannot be dealt with satisfactorily in a section such as this—those interested should see Cellular-news (2012a) and Heavy Reading (2011). It is worth noting, however, that several operators have recently gone public with their plans to build such networks. South Korea, inevitably, will probably be in the forefront of small cell development, but Sprint Nextel is also leading the way using the LightRadio architecture developed by Alcatel-Lucent. The idea is to begin indoors and progress outdoors with hotzones of capacity where the network is under particular pressure. The architecture is designed to cope with all generations of mobile technology as well as Wi-Fi, although in this case Sprint Nextel intends to use it exclusively to augment its new LTE network (Gabriel 2012e).

1.15   Satellite Provision

It has been necessary, in order to prepare a cohesive case study, to include a detailed discussion of satellite provision in the USA within the relevant section below, and for that reason this section is confined to the European Union. As the USA case study demonstrates, satellite provision is straightforward in principle but very difficult to implement in practice, but progress has been made and three companies are actively involved as well as terrestrial counter-parties.

As for the EU, the picture is somewhat depressing. Satellite provision was built into the plans for 3G as far back as 1997 (Curwen 2002: pp. 30–38), but the industry already had a long history of failure and bankruptcy to contend with. In June 1997, the European Radiocommunications Committee decided (ERC/DEC/(97)07) to allocate 1,980–2,010 MHz paired with 2,170–2,200 MHz for satellite applications using a space division multiple access (SDMA) interface. This eventually became Decision No. 2007/98/EC of the European Commission and was essentially concerned with spectrum usage conditions. A further Decision (No. 626/2008/EC) of the European Parliament and the Council in June 2008 was concerned with the selection and authorisation of systems providing mobile satellite services (MSS). It provided a legal framework for a two-phase selection process and a Commission Decision on the selection of operators (No. 2009/449/EC) was adopted on 13 May 2009, in the process selecting Inmarsat Ventures and Solaris Mobile.

The next step was intended to be the granting of authorisations at the national level. Should the authorised party fail to deliver in any individual case, a procedure would be set in hand which would initially not include withdrawal of the authorisation although that was provided for as the ultimate sanction. The licensees were given 2 years to launch MSS across Europe. This they failed to do and in October 2011, the procedures to deal with the problem were initiated. At this point in time, Solaris had been granted authorisations in 16 member states but had made little real progress towards a launch anywhere other than in Italy. On the other hand, as is evident, a significant number of member states had yet even to complete the authorisation process, so it is fair to argue that leaving licensing of satellite provision to national governments in Europe is not a recipe for success, either in the short term or, possibly, at all. Sooner or later, therefore, the issue of whether to re-farm the band must be addressed.

Meanwhile, there are signs of progress in relation to the L-band (1,452–1,492 MHz), which is also significant in the US context. Among other technical developments, Supplemental Downlink (SDL) is being developed as a means whereby operators can make use of fragmented spectrum holdings. SDL is a form of Carrier Aggregation which enables TDD spectrum in the likes of (especially) the 700 MHz band to be combined with the L-band to provide supplemental capacity for a FDD service in a different band. In Europe, the CEPT authorised the use of the L-band for FDD in 2012 by way of a decision that both harmonised the band and reserved it for SDL. Almost immediately, in June 2012, Orange in France obtained a licence to test the technology in Toulouse in conjunction with Qualcomm and Ericsson using a combination of L-band and 3G frequencies, and the first fruits of this were revealed in February 2013 (White 2013b).

1.16 Network Sharing

Reference to a number of network sharing arrangements is contained in the case studies below. The purpose of this section is to examine the issue across the board. There is nothing new about this topic which was regularly discussed in the context of 3G, but the financial problems that beset a significant number of 3G licensees when attempting to roll out their networks in isolation has meant that there has been far greater readiness to address, and indeed to resolve, this issue in the context of LTE.

There are obvious reasons to justify network sharing: it reduces costs for each partner, speeds up network coverage and reduces environmental side-effects. At the end of the day, if a common network has sufficient capacity to serve most or all of the operators, it is seemingly a 'no-brainer' to build it. However, there is an inevitable fly in the ointment, namely how to maintain inter-operator competition.

To understand this it is first necessary to distinguish between 'passive' and 'active' elements of a network. The passive elements are primarily the sites themselves, the masts and antennae erected on them and the power supplies and other equipment need for their optimum functioning. The active elements include the radio access network (RAN) including base stations, radio network controllers and backhaul. It is immediately evident that sharing passive elements is much less of a concern that sharing active elements. At the end of the day, operators need to distinguish the services that they sell, and if the signal to the consumer arrives along an entirely common route then competition may well end up in the form of cut-throat pricing, which is clearly not the object of the exercise since not all operators will survive.

It is also worth bearing in mind that MVNOs have to use the infrastructure of the network that hosts them. A competitive market must therefore provide for the requisite spare capacity on such networks as exist and must not allow the network operator to charge fees that will make it impossible for MVNOs to compete.

This was an issue when the joint venture between TeliaSonera and Telenor in Denmark, known as TT-Netværket, was set up. Eventually, a list of conditions was drawn up that was satisfactory to both the two operators and the regulator but this will obviously have to be rigorously policed over time.

1.17 Cognitive Radio

Cognitive radio is a term used in the context of spectrum sharing—for a not overly-technical introduction see Wikipedia (2013b) which also considers it in comparison to intelligent antenna and for a highly detailed discussion of its many facets see the special issue of *Telecommunications Policy* on "Cognitive radio and dynamic spectrum assignment" (Volume 37, Issues 2–3, March/April 2013). In effect, base stations and devices linking to them search out frequencies that are not in use at that point in time, and put them temporarily to use in making a connection. Generally, when one operator controls a spectrum band, no matter how narrow, it has the choice as to whether or not to use it intensively (or, indeed, at all). Cognitive radio means that lots of unrelated operations can use the same band, dipping in and out whenever it is free, and hence the intensity of use is very likely to be greater.

It is important to note that there are no plans currently to use cognitive radio in licensed spectrum set aside for mobile networks. Rather, there is an acknowledge-ment that spectrum held by the military and a variety of public bodies is often used inefficiently and that cognitive radio may provide a solution that reduces waste of a scarce resource. Unfortunately, despite years of research, little progress has been made in setting out the necessary rules and agreements that will need to be adopted if a free-for-all approach is to be avoided and with it the interference problems that often blight licence-exempt frequencies. However, this should not come as too big a surprise as most bodies that have secured spectrum rights are notoriously reluctant to make any concessions without an attractive quid-pro-quo (which may render the whole exercise economically pointless).

In recent times—the matter is occasionally referred to in the case studies in the chapter that follows—cognitive radio has become associated with the use of TV White Spaces (TVWS) such that standard-setting bodies, the likes of the ITU and country regulators increasingly see the use of TVWS as the testing ground for the feasibility of cognitive radio. Unfortunately (once again) an ideal solution presupposes that every country is switching from analogue to digital provision at roughly the same time and that every one intends to use precisely the same frequencies to provide digital TV, which is far from being the reality. Furthermore, interference issues are particularly thorny to resolve along the borders between countries where there is absolutely no consistent model suitable for all-comers. Less contentiously there are issues as to whether, for example, to apply cognitive radio across an entire country or merely in urban or rural areas depending on reception quality.

References

3G Americas. (2006). *Mobile broadband: The global evolution of UMTS/HSPA release 7 and beyond.* Accessed August 17, 2010, from http://www.3gamericas.com

3G Americas. (2007). *UMTS Evolution from 3GPP Release 7 to Release 8 HSPA and SAE/LTE.* Accessed October 13, 2010, from http://www.3gamericas.com

3G.co.uk. (2009a). *NTT DoCoMo's super 3G (LTE).* Accessed July 23, 2010, from http://www.3g.co.uk

3G.co.uk. (2009b). *HSPA+ solution to double 3G download speeds.* Accessed September 23, 2010, from http://www.3g.co.uk

4G Americas. (2011). *The benefits of using LTE in digital dividend spectrum,* October. http://www.4gamericas.com

4G Americas. (2012a). *HSPA+ LTE carrier Aggregation,* June. http://www.4gamericas.com

4G Americas. (2012b). *MIMO and smart antennas for mobile broadband systems,* October. http://www.4gamericas.com

4G-Portal. (2013a). *SK Telecom and Ericsson on the way towards LTE-A.* Accessed January 31, 2013, from http://www.4g-portal.com

4G-Portal. (2013b). *ZTE and Imagination Technologies announce availability of VoLTE on commercial handsets.* Accessed February 27, 2013, from http://www.4g-portal.com

Aetha. (2011a). *Case studies for the award of the 700MHz/800MHz band: Introduction.* Accessed November 20, 2011, from http://www.gsma.com/spectrum

Aetha. (2011b). *Case studies for the award of the 700MHz/800MHz band: Germany.* Accessed November 20, 2011, from http://www.gsma.com/spectrum

Aetha. (2011c). *Case studies for the award of the 700MHz/800MHz band: Mexico.* Accessed November 20, 2011, from http://www.gsma.com/spectrum

Aetha. (2011d). *Case studies for the award of the 700MHz/800MHz band: Finland.* Accessed November 20, 2011, from http://www.gsma.com/spectrum

Aetha. (2011e). *Case studies for the award of the 700MHz/800MHz band: Australia.* Accessed November 20, 2011, from http://www.gsma.com/spectrum

Aetha. (2012, July). *Spectrum value of 800MHz, 1800MHz and 2.6GHz: A DotEcon and Aetha Report for Ofcom.* DotEcon, London.

Arora, N. (2013). *Carrier aggregation in LTE-advanced – whitepaper.* Accessed March 19, 2013, from http://www.4g-portal.com

Cellular-news. (2010). *Huawei unveils world's first triple-mode LTE modem.* Accessed February 17, 2010, from http://www.cellular-news.com

Cellular-news. (2011a). *Global Certification Forum starts working with TD-LTE handsets.* Accessed July 11, 2011, from http://www.cellular-news.com

Cellular-news. (2011b). *ZTE participates in first-ever MSF VoLTE interoperability test.* Accessed October 25, 2011, from http://www.cellular-news.com

Cellular-news. (2011c). *HSPA technology standards could deliver up to 336 Mbps speeds.* Accessed October 26, 2011, from http://www.cellular-news.com

Cellular-news. (2012a). *Small cells loom large for LTE rollouts.* Accessed June 6, 2012, from http://www.cellular-news.com

Cellular-news. (2012b). *VoLTE service revenues predicted to reach $2 billion by 2016.* Accessed June 25, 2012, from http://www.cellular-news.com

Cellular-news. (2013a). *Telefonica demos voice call handover from VoLTE to 3G network.* Accessed February 7, 2013, from http://www.cellular-news.com

Cellular-news. (2013b). *GTI members announce successful TD-LTE/LTE FDD global roaming trials.* Accessed February 26, 2013, from http://www.cellular-news.com

Cellular-news. (2013c). *EC announces EUR 50 mln funds for 5G mobile research.* Accessed February 2013, from http://www.cellular-news.com

Cellular-news. (2013d). *Next generation VoLTE smartphones offer significant battery life improvement.* Accessed March 13, 2013, from http://www.cellular-news.com

Curwen, P. (2002). *The future of mobile communications: Awaiting the third generation.* Basingstoke: Palgrave Macmillan.

Dahlman, E., Parkvall, S., & Skold, J. (2011). *4G: LTE/LTE advanced for mobile broadband.* Oxford: Academic.

Dailywireless. (2010). *WiMAX2 collaboration initiative.* Accessed December 15, 2010, from http://www.dailywireless.org

Dow Jones. (2011). *China won't launch commercial 4G services until 2014 – Minister.* Accessed March 12, 2011, from http://www.totaltele.com

Ericsson. (2013). *LTE Release 12 – Taking another Step toward the Networked Society. White paper 284-23-3189 Uen.* Accessed January 15, 2013, from http://www.ericsson.com

Feasey, R. (2012, May). Primary reassignments. *The Policy Paper Series,* 14, pp. 3–6.

Federal Communications Commission. (2007). *Auction of 700 MHz band licences FCC Public Notice DA 07–4514 of 2 November.*

Gabriel, C. (2010a). *WiMAX giants step up defense against LTE.* Accessed April 13, 2010, from http://www.rethink-wireless.com

Gabriel, C. (2010b). *Six operators aim for LTE bloc in 450MHz.* Accessed September 27, 2010, from http://www.rethink-wireless.com

Gabriel, C. (2010c). *Samsung demonstrates WiMAX2 at 330Mbps.* Accessed October 5, 2010, from http://www.rethink-wireless.com

Gabriel, C. (2010d). *ITU picks WiMAX2 and LTE-Advanced for 'true 4G'.* Accessed October 25, 2010, from http://www.rethink-wireless.com

Gabriel, C. (2011a). *First true 4G standard approved with 802.16m.* Accessed April 10, 2011, from http://www.rethink-wireless.com

Gabriel, C. (2011b). *Metro PCS in aggressive LTE voice plan.* Accessed August 6, 2011, from http://www.rethink-wireless.com

Gabriel, C. (2012a). *ZTE and Huawei vye for LTE-A firsts.* Accessed February 29, 2012, from http://www.rethink-wireless.com

Gabriel, C. (2012b). *Huawei claims 30 Gbps 'Beyond LTE'.* Accessed March 21, 2012, from http://www.rethink-wireless.com

Gabriel, C. (2012c). *ZTE pushes 3.5GHz as LTE band.* Accessed March 22, 2012, from http://www.rethink-wireless.com

Gabriel, C. (2012d). *Huawei leads charge towards LTE-advanced.* Accessed May 14, 2012, from http://www.rethink-wireless.com

Gabriel, C. (2012e). *Sprint to use lightRadio for LTE.* Accessed August 6, 2012, from http://www.rethink-wireless.com

Gabriel, C. (2012f). *China and Taiwan get closer on 4G.* Accessed September 3, 2012, from http://www.rethink-wireless.com

Gabriel, C. (2012g). *WiMAX opens arms to TD-LTE at last.* Accessed October 31, 2012, from http://www.rethink-wireless.com

Gabriel, C. (2013a). *Wi-Fi will be embedded in AT&T's small cell plan.* Accessed January 10, 2013, from http://www.rethink-wireless.com

Gabriel, C. (2013b). *Samsung claims '5G' breakthrough.* Accessed May 13, 2013, from http://www.rethink-wireless.com

Gabriel, C. (2013c). *SKT and Samsung team on 'LTE-A' launch.* Accessed June 26, 2013, from http://www.rethink-wireless.com

Global Insight. (2007a). *WiMAX gains initial approval as IMT-2000 3G standard.* Accessed June 6, 2009, from https://www.communicationsdirectnews.com

Global Insight. (2007b). *WiMAX as part of 3G – a boost or a step back?* Accessed October 23, 2009, from https://www.communicationsdirectnews.com

Global Insight. (2009). *Finland allocates new 4G frequencies to TeliaSonera, Elisa, DNA.* Accessed April 25, 2010, from https://www.communicationsdirectnews.com

Global mobile Suppliers Association. (2010a). *GSM/3G market/technology update.* Accessed November 1, 2010, from http://www.gsacom.com

Global mobile Suppliers Association. (2010b). *Digital dividend update*. Accessed April 1, 2011, from http://www.gsacom.com

Global mobile Suppliers Association. (2012a). *Mobile HD voice: Global update report*. Accessed September 23, 2012, from http://www.gsacom.com

Global mobile Suppliers Association. (2012b). *Status of the global LTE 1800 market*. Accessed November 14, 2012, from http://www.gsacom.com

Global mobile Suppliers Association. (2012c). *Evolution to LTE report*. Accessed September 13, 2012, from http://www.gsacom.com

Global mobile Suppliers Association. (2012d). *Status of the global LTE TDD market*. Accessed November 22, 2012, from http://www.gsacom.com

GSM World. (2010). *GSMA VoLTE initiative*. Accessed December 22, 2010, from http://www.gsmworld.com

Heavy Reading. (2011). Wi-Fi strategies for mobile operators. *Heavy Reading, 9*, 10.

Hetting, C., & Stanislawski, S. (2010). *LTE TDD: The preferred choice for mobile broadband in unpaired bands*. A report published by Ventura Team in March 2010.

Hibberd M. (2013). *EC grants 800MHz delays to nine member states*. Accessed July 23, 2013, from http://www.telecoms.com

Irmer, R., Droste, H., Marsch, P., Grieger, M., Fettweis, G., Brueck, S., et al. (2011, February). Coordinated multipoint: Concepts, performance and field trial results. *IEEE Communications Magazine*, 102–111.

Khan, A. (2012). *VoLTE subscriptions to hit 2 million by 2013*. Accessed May 16, 2012, from http://www.cellular-news.com

Kimura, D., & Seki, H. (2012). Inter-cell interference coordination technology. *Fujitsu Scientific and Technical Journal, 48*(1), 89–94.

Lteportal. (2011). *The LTE/LTE-advanced guide*. Accessed March 2, 2011, from http://www.lteportal.com

Middleton, J. (2012). *LTE TDD winning wide acceptance*. Accessed June 25, 2012, from http://www.telecoms.com

Middleton, J. (2013). *A new lease of life*. Accessed January 4, 2013, from http://www.telecoms.com

Morris, A. (2010). *True 4G specs to be released in March 2011*. Accessed February 18, 2010, from http://www.totaltele.com

Poikselkä, M., Holma, H., Hongisto, J., Kallio, J., Toskala, A., & Ebooks Corporation. (2012). *Voice over LTE (VoLTE)*. Chichester: Wiley.

Sahota, D. (2013). *NSN claims record upload TD-LTE speeds*. Accessed June 25, 2013, from http://www.telecoms.com

Sesia, S., Toufik, I., & Baker, M. (2011). *LTE – the UMTS long term evolution: From theory to practice*. Chichester: Wiley.

Technology Marketing Corporation. (2011). *NextGen voice*. Accessed June 14, 2011, from http://www.tmcnet.com

Telecom.paper. (2011a). *EU opens up GSM bands for 4G services*. Accessed April 20, 2011, from http://www.telecompaper.com

Telecom.paper. (2011b). *Ericsson demonstrates LTE advanced in Sweden*. Accessed June 29, 2011, from http://www.telecompaper.com

Telecom.paper. (2011c). *MSN to demonstrate 336 Mbps HSPA+ data call*. Accessed September 27, 2011, from http://www.telecompaper.com

Telecom.paper. (2012a). *China Mobile, Clearwire, agree on common TD-LTE testing*. Accessed January 18, 2012, from http://www.telecompaper.com

Telecom.paper. (2012b). *IMT-advanced standards agreed at ITU assembly*. Accessed January 19, 2012, from http://www.telecompaper.com

Telecom.paper. (2013a). *Movistar, Vodafone, Orange launch Joyn-embedded handsets*. Accessed April 2, 2013, from http://www.telecompaper.com

Telecom.paper. (2013b). *Huawei Touch complete LTE carrier aggregation field trial.* Accessed April 8, 2013, from http://www.telecompaper.com

Telecom.paper. (2013c). *Broadcom, SK Telecom launch 5G Wi-Fi router.* Accessed April 8, 2013, from http://www.telecompaper.com

TeleGeography. (2012a). *LTE advanced tests reported by Tele2, Telenor.* Accessed May 14, 2012, from http://www.telegeography.com

TeleGeography. (2012b). *Ericsson performs first live FDD/TDD handover on China Mobile HK's network.* Accessed June 22, 2012, from http://www.telegeography.com

TeleGeography. (2012c). *MetroPCS expects to launch VoLTE in all markets within six months.* Accessed September 21, 2012, from http://www.telegeography.com

TeleGeography. (2012d). *Ericsson and China Mobile showcase 223Mbps TDD-LTE.* Accessed November 8, 2012, from http://www.telegeography.com

TeleGeography. (2012e). *Tele2 'first in Europe' to test VoLTE.* Accessed December 21, 2012, from http://www.telegeography.com

TeleGeography. (2013). *SK Telecom inaugurates commercial LTE-A network.* Accessed June 26, 2013, from http://www.telegeography.com

Vesterdorf, B. (2012, May). Radio spectrum licence renewals. *The Policy Paper Series*, 14, pp. 16–24.

Weaver, P. (2011). *Everyone's first with VoLTE.* Accessed June 15, 2011, from http://www.telecoms.com

White, P. (2013a). *Cisco and NSN meet at the Wi-Fi/cellular border.* Accessed February 26, 2013, from http://www.rethink-wireless.com

White, P. (2013b). *Orange demonstrates technique to use L-band spectrum.* Accessed February 26, 2013, from http://www.rethink-wireless.com

Wikipedia. (2010a). *Evolved HSPA.* Accessed June 1, 2010, from http://www.en.wikipedia.org/wiki/HSPA%2B

Wikipedia. (2010b). *Evolution-Data Optimized.* Accessed June 1, 2010, from http://www.en.wikipedia.org/wiki/EVDO

Wikipedia. (2010c). *High Speed Packet Access.* Accessed June 1, 2010, from http://www.en.wikipedia.org/wiki/High_Speed_Packet_Access

Wikipedia. (2010d). *WiMAX.* Accessed June 1, 2010, from http://www.en.wikipedia.org/wiki/Wimax

Wikipedia. (2010e). *3GPP.* Accessed June 1, 2010, from http://www.en.wikipedia.org/wiki/3GPP

Wikipedia. (2010f). *3GPP Long Term Evolution.* Accessed June 1, 2010, from http://www.en.wikipedia.org/wiki/3GPP_Long_Term_Evolution

Wikipedia. (2010g). *4G.* Accessed December 1, 2010, from http://www.en.wikipedia.org/wiki/4G

Wikipedia. (2012a). *Multi-user MIMO.* Accessed September 17, 2012, from http://www.en.wikipedia.org/wiki/Multi-user_MIMO

Wikipedia. (2012b). *Wideband audio.* Accessed September 26, 2012, from http://www.en.wikipedia.org/wiki/Wideband_audio

Wikipedia. (2013a). *Rich Communication Services.* Accessed January 12, 2013, from http://www.en.wikipedia.org/wiki/Rich_Communication_Services

Wikipedia. (2013b). *Cognitive radio.* Accessed January 12, 2013, from http://www.en.wikipedia.org/wiki/Cognitive_radio

WiMAX Forum. (2009). *WiMAX Forum overview.* Accessed June 1, 2010, from http://www.wimaxforum.org/about

LTE Case Studies Overview

2

2.1 Introduction

The case studies which follow are based upon information supplied by a wide variety of internet websites, all of which have been cross-checked. There is only one alternative source for this information in the public domain which is the Global mobile Suppliers Association (GSA) at http://www.gsacom.com. However, it uses a somewhat different system for recording LTE launches. For example, it identifies many of the networks differently, it includes certain of the launches itemised as excluded in footnote 1 to Table 2.1– in particular wholesalers (see Global mobile Suppliers Association 2012)—and its case studies are prepared independently from those below and hence differ considerably in their emphasis and level of detail—for example, they do not contain detailed information about licensing—in addition to the fact that they are not written as a continuous narrative. It may be added that there is (inevitably) ambiguity about the definition of a launch. For the purposes of Table 2.1, launch dates are (insofar as it can be ascertained) those when a network goes 'commercial' even though the area/population covered may be quite small. If the launch is explicitly stated to be a 'trial', 'test', 'pilot' or equivalent, then that is simply noted in the table and case studies.

Overall, the above clearly raises the need to warn readers about the interpretation of any aggregated data that are reported in the media. In particular, the number of declared launches can be altered significantly by variations in the treatment of regional operators, MVNOs and wholesalers. Hopefully, Table 2.1 is set out in such a way as to allow readers to make any adjustments for themselves.

Readers are also strongly advised that second-hand versions of the GSA database are not to be trusted. For example, that available at http://www.teleco.com, the major mobile telecommunications site for Latin America, has at various times introduced numerous errors including the following: MTS (Armenia) listed as launching in both February 2011 and December 2011; Yota Belarus listed as launching in both December 2011 and May 2012; Glo Mobile (Niger) listed as launching in January 2011 although it does not exist—it exists in Nigeria but has

P. Curwen and J. Whalley, *Fourth Generation Mobile Communication*,
Management for Professionals, DOI 10.1007/978-3-319-02210-9_2,
© Springer International Publishing Switzerland 2013

Table 2.1 LTE launches and commitments.[a] June 2013

Country	Operator	MHz Band[b]	Launch	Comment
Afghanistan	Etisalat	–	–	*See case study Chap. 6*
Algeria	–	–	–	*See case study Chap. 7*
Andorra	Andorra Telecom	2,600	–	–
Angola	Movicel	1,800	04/12	(See TeleGeography 2012b)
Angola	Unitel	2,100	12/12	(See Engineering news 2013)
Anguilla	Digicel	–	–	Marketing HSPA + as '4G'
Antigua	Digicel	700	11/12	*See case study Chap. 7*
Argentina	Personal	n/a	–	*See case study Chap. 7*
Argentina	Telefónica	AWS	2013	
Armenia	MTS	2,600	01/12	Awarded spectrum 11/10[c]
Armenia	VimpelCom	698–806	–	Based on tests in Kazakhstan
Armenia	Orange	n/a	–	
Australia	Optus	1,800/2,300	07/12	*See case study Chap. 6*
Australia	Telstra	1,800/2,600	09/11	
Australia	VHA	1,800	06/13	
Austria	Hutchison 3G	2,600	11/11	*See case study Chap. 5*
Austria	Orange	2,600	–	
Austria	Telekom Austria	2,600	11/10	
Austria	T-Mobile	2,600	07/11	
Azerbaijan	Azerfon	n/a	–	*See case study Chap. 7*
Azerbaijan	Bakcell	n/a	2013	
Azerbaijan	TeliaSonera	1,800	05/12	
Bahamas	BTC	700	–	*See case study Chap. 7*
Bahamas	C&W	700	–	
Bahrain	Zain	n/a	04/13	*See case study Chap. 7*
Bahrain	Batelco	1,800	02/13	
Bahrain	STC	1,800	–	
Bangladesh	–	–	–	*See case study Chap. 6*
Belarus	Turkcell	2,600	–	*See case study Chap. 7*
Belarus	MTS	2,600	–	
Belgium	Belgacom	1,800/2,600	11/12	*See case study Chap. 5*
Belgium	KPN	2,600	–	
Belgium	Orange	1,800/2,600	–	
Belgium	BUCD	2,600	–	
Belize	Belize Telecom	Existing	–	–
Benin	MTN	n/a	–	3G/4G licence awarded in March 2012
Bhutan	Bhutan telecom	1,800	2013	*See case study Chap. 6*
Bolivia	Entel	700	12/12	*See case study Chap. 7*

(continued)

Table 2.1 (continued)

Country	Operator	MHz Band[b]	Launch	Comment
Bosnia & Herz.	Telekom Srpske	n/a	–	*See case study Chap. 7*
Bosnia & Herz.	BH Telekom	–	–	
Botswana	MTN (Mascom)	n/a	–	*See case study Chap. 7*
Brazil	Oi	2,600	04/13	*See case study Chap. 7*
Brazil	Telefónica	2,600	04/13	
Brazil	América Móvil	2,600	12/12	
Brazil	TIM	2,600	07/13	
Br. Virgin Isles	–	–	–	*See case study Chap. 7*
Brunei	–	1,800	–	*See case study Chap. 6*
Bulgaria	Telekom Austria	1,800	01/12	*See case study Chap. 5*
Bulgaria	Vivacom	–	–	
Bulgaria	Max Telecom	1,800	2013	
Bulgaria	Bulsatcom	1,800	2013	
Cambodia	Digital Star Media	n/a	2013	Plans launch by acquiring excell
Canada	Bell Canada	AWS/2,600	09/11	*See case study Chap. 7*
Canada	Rogers	AWS/2,600	07/11	
Canada	Telus	AWS	02/12	
Chile	América Móvil	2,500	01/13	*See case study Chap. 7*
Chile	Telefónica	2,500	2013	
Chile	Entel	2,500	2013	
China	China Mobile	2,570–2,620	2014	*See case study Chap. 6*
China	China Telecom	n/a	2014	
Colombia	UNE-EPM	2,600	06/12	*See case study Chap. 7*
Colombia	América Móvil	1,900	–	
Colombia	Telefónica	1,900	–	
Colombia	Millicom	1,900	–	
Costa Rica	ICE	2,600	06/13	*See case study Chap. 7*
Costa Rica	América Móvil	2,600	–	
Croatia	T-Mobile	1,800/800	03/12	*See case study Chap. 5*
Croatia	Telekom Austria	1,800	03/12	
Cyprus South	–	–	–	*See case study Chap. 5*
Czech Rep.	Telefónica	1,800	06/12	*See case study Chap. 5*
Czech Rep.	T-Mobile	Multiple	2013	
Czech Rep.	Vodafone	Multiple	2013	
Denmark	TDC	2,600/800	10/11	*See case study Chap. 5*
Denmark	Telenor	1,800/2,600	2013	

(continued)

Table 2.1 (continued)

Country	Operator	MHz Band[b]	Launch	Comment
Denmark	TeliaSonera	Multiple	12/10	
Denmark	Hutchison '3'	1,800/2,600	09/12	
Djibouti	Djibouti Telecom	n/a	01/13	In practice, HSPA + (TeleGeography 2013a)
Dominican Rep.	Orange	1,800	01/13	*See case study Chap. 7*
Dominican Rep.	Tricom	800/1,900	03/13	
DR Congo	Smile Comms	790–862	–	
DR Congo	Millicom	–	05/13	In practice HSPA + (TeleGeography 2013d)
Ecuador	América Móvil	n/a	–	*See case study Chap. 7*
Ecuador	CNT	700/AWS	–	
Egypt	Vodafone	Existing	–	Testing
Egypt	MobiNil	Existing	–	Testing
Egypt	Etisalat	Existing	–	Testing
Estonia	Elisa	Multiple	02/13	*See case study Chap. 5*
Estonia	Tele2	2,600/1,800	11/12	
Estonia	TeliaSonera	2,600/1,800	12/10	
Ethiopia	Ethio Telecom	n/a	–	Rolling out network (Cellular-news 2012b)
Fiji	Fiji Telecom	–	–	*See case study Chap. 6*
Fiji	Digicel	–	–	
Fiji	Vodafone	–	–	
Finland	DNA	2,600/1,800	12/11	*See case study Chap. 5*
Finland	Elisa	2,600/1,800	12/10	
Finland	TeliaSonera	2,600/1,800	11/10	
France	Orange	800/2,600	11/12	*See case study Chap. 5*
France	Vivendi	800/2,600	11/12	
France	Bouygues	800/2,600	05/13	
France	Iliad	800/2,600	–	
Fr. Polynesia	–	–	–	Vendors testing feasibility of 4G
Georgia	VimpelCom	1,800	–	Testing commenced November 2011
Germany	KPN	2,600/1,800	–	*See case study Chap. 5*
Germany	Telefónica	800/2,600	07/11	
Germany	T-Mobile	Multiple	12/10	
Germany	Vodafone	800/2,600	09/10	
Ghana	Glo Mobile	n/a	–	*See case study Chap. 7*
Greece	OTE	900/1,800	11/12	*See case study Chap. 5*
Greece	Vodafone	900/1,800	12/12[d]	
Greece	Wind	900/1,800	–	
Greenland	TELE Greenland	800	2012Q4	*See case study Chap. 7*
Guam	IT&E	700	08/12	*See case study Chap. 6*

(continued)

Table 2.1 (continued)

Country	Operator	MHz Band[b]	Launch	Comment
Guam	DoCoMo	700	10/12	
Guam	iConnect	700	03/13	
Guatemala	–	Multiple	–	See case study Chap. 7
Haiti	Digicel	–	–	See case study Chap. 7
Hong Kong	CSL	1,800/2,600	11/10	See case study Chap. 6
Hong Kong	Hutchison	1,800/2,600	05/12	
Hong Kong	PCCW	1,800/2,600	04/12	
Hong Kong	China Mobile	2,600/2,300	04/12	
Hong Kong	SmarTone	1,800	08/12	
Hungary	Vodafone	900/2,600	2013	See case study Chap. 5
Hungary	Telenor	1,800/2,600	02/13	
Hungary	T-Mobile	1,800/900	01/12	
Iceland	Novator	1,800/800	04/13	See case study Chap. 5
India	Aircel	2,300	2013	See case study Chap. 4
India	Bharti Airtel	2,300	04/12	
India	BSNL	2,300	–	
India	MTNL	2,300	–	
India	Qualcomm	2,300	2013	
India	Reliance Ind.	2,300	2013	
India	Tikona Digital	2,300	2013	
India	Videocon	1,800	2013	
Indonesia	Axiata	–	–	See case study Chap. 6
Indonesia	Indosat	1,800	–	
Indonesia	Telkomsel	–	–	
Iraq	–	–	–	See case study Chap. 7
Ireland	–	2,600	–	See case study Chap. 5
Isle of Man	Manx Telecom	2,600	–	Trials from November 2011
Israel	–	–	–	See case study Chap. 7
Italy	Telecom Italia	1,800/2,600	11/12	See case study Chap. 5
Italy	Vodafone	1,800/2,600	10/12	
Italy	Wind	1,800/2,600	2013	
Italy	Hutchison	1,800/2,600	11/12	
Jamaica	–	700	–	See case study Chap. 7
Japan	DoCoMo	2,100/1,500	12/10	See case study Chap. 6
Japan	KDDI	Multiple	09/12	
Japan	eMobile	1,700	03/12	
Japan	SoftBank Mobile	Multiple	02/12[e]	
Jersey	Clear Mobitel	2,600	–	See case study Chap. 5
Jordan	Zain	Existing	–	See case study Chap. 7
Kazakhstan	VimpelCom	700–799	–	See case study Chap. 7
Kazakhstan	Kazakhtelecom	1,800	12/12	
Kenya	Various	n/a	–	See case study Chap. 7

(continued)

Table 2.1 (continued)

Country	Operator	MHz Band[b]	Launch	Comment
Kiribati	TSKL	700	–	*See case study Chap. 6*
Kuwait	Viva	1,800	12/11	*See case study Chap. 7*
Kuwait	Zain	1,800	11/12	
Kuwait	Qatar Telecom[f]	1,800	07/13	
Kyrgyzstan	Saima Telecom	2,600	12/11	*See case study Chap. 7*
Laos	LTC	n/a	–	*See case study Chap. 6*
Laos	VimpelCom	1,800/2,600	–	
Latvia	Tele2	2,600	2013	*See case study Chap. 5*
Latvia	TeliaSonera	1,800/2,600	05/11	
Latvia	Bité	2,600	–	
Latvia	Baltkom	2,600	–	
Latvia	Triatel	790–862	–	
Lebanon	Alfa Telecom	800/2,600	05/13	*See case study Chap. 7*
Lebanon	MTC	800/2,600	05/13	
Liberia	Cellcom	n/a	–	*See case study Chap. 7*
Libya	Al Madar	–	–	Government mulling over licence issue
Libya	Libyana	–	–	
Lithuania	TeliaSonera	1,800	05/11	*See case study Chap. 5*
Lithuania	Bité	2,600	–	
Lithuania	Tele2	2,600	03/13	
Luxembourg	Orange	1,800	11/12	*See case study Chap. 5*
Luxembourg	Belgacom	1,800	10/12	
Macedonia	T-Mobile	800/1,800	–	*See case study Chap. 5*
Macedonia	Telekom Austria	800/1,800	–	
Macedonia	Telekom Slovenije	800/1,800	–	
Malaysia	Maxis	2,600	01/13	*See case study Chap. 6*
Malaysia	Axiata	2,600	04/13	
Malaysia	DiGi	2,600	07/13	
Maldives	Qatar Telecom[f]	–	2013	On trial
Mauritius	Orange	1,800	06/12	*See case study Chap. 7*
Mauritius	Millicom	1,800	05/12	
Mexico	América Móvil	AWS	11/12	*See case study Chap. 7*
Mexico	Telefónica	AWS	11/12	
Moldova[g]	Moldtelecom	2,600/3,700	–	*See case study Chap. 7*
Moldova	Orange	2,600/3,700	11/12	
Moldova	TeliaSonera	2,600/3,700	11/12	
Monaco	Monaco Telecom	2,600	2013	*See case study Chap. 5*
Montenegro	Telenor	2,600	11/12	*See case study Chap. 5*
Morocco	–	n/a	2013	*See case study Chap. 7*
Mozambique	–	–	–	*See case study Chap. 7*

(continued)

Table 2.1 (continued)

Country	Operator	MHz Band[b]	Launch	Comment
Myanmar	–	–	–	*See case study Chap. 6*
Namibia	MTC	1,800	05/12	*See case study Chap. 7*
Nepal	Ncell	n/a	–	Has requested licence and spectrum from regulator
Nepal	Nepal Telecom	2,300	–	TD-LTE
Netherlands	KPN	2,600/800	05/12[d]	*See case study Chap. 5*
Netherlands	T-Mobile	2,600/1,800	05/12[d]	
Netherlands	Vodafone	2,600	05/12[d]	
New Zealand	TNZ	698–806	–	*See case study Chap. 6*
New Zealand	Vodafone	1,800	02/13	
Nicaragua	–	–	–	*See case study Chap. 7*
Nigeria	Globacom	Existing	–	*See case study Chap. 7*
Norway	Arctic Wireless	2,600	–	*See case study Chap. 5*
Norway	Craig Wireless	2,600	–	
Norway	Telenor	1,800/2,600	10/12	
Norway	TeliaSonera	1,800/2,600	12/09	
Oman	Omantel	1,800/2,300	–	*See case study Chap. 7*
Oman	Qatar Telecom[f]	1,800	02/13	
Pakistan	–	–	–	*See case study Chap. 6*
Paraguay	Millicom	1,900	–	*See case study Chap. 7*
Paraguay	Copaco	AWS	02/13	
Paraguay	Personal	1,900	02/13	
Peru	Telefónica	n/a	–	*See case study Chap. 7*
Peru	Entel	AWS	–	
Peru	América Móvil	n/a	–	
Philippines	Globe	1,800	09/12	*See case study Chap. 6*
Philippines	Smart	Multiple	08/12	
Poland	Aero2 (Mobyland)	1,800	09/10	*See case study Chap. 5*
Poland	Orange/Netia P4	–	–	
Poland	Polkomtel	900/2,600	12/11	(as MVNO)
Poland	T-Mobile	–	–	
Portugal	Portugal Telecom	Multiple	03/12	*See case study Chap. 5*
Portugal	Vodafone	Multiple	03/12	
Portugal	Sonae.com	Multiple	03/12	
Puerto Rico	América Móvil	700	11/12	*See case study Chap. 7*
Puerto Rico	AT&T	700/AWS	11/11	
Puerto Rico	Open Mobile	700	04/12	
Puerto Rico	Sprint Nextel	1,900	12/12	
Puerto Rico	T-Mobile	700	07/13	
Qatar	Qatar Telecom[f]	800/2,600	04/13	*See case study Chap. 7*

(continued)

Table 2.1   (continued)

Country	Operator	MHz Band[b]	Launch	Comment
Qatar	Vodafone	Existing	–	
Romania	Orange	1,800	12/12	*See case study Chap. 5*
Romania	OTE	Multiple	04/13	
Romania	Vodafone	1,800	11/12	
Romania	RCS&RDS	Multiple	–	
Russia	MegaFon	800/2,600	06/13	*See case study Chap. 4*
Russia	MTS	800/2,600	09/12	
Russia	VimpelCom	800/2,600	05/13	
Rwanda	–	–	–	*See case study Chap. 7*
Samoa	Digicel	n/a	03/12	Probably HSPA+
Saudi Arabia	Etisalat	2,600	09/11	*See case study Chap. 7*
Saudi Arabia	STC	2,300–2,390	09/11	
Saudi Arabia	Zain	1,800	09/11	
Singapore	MobileOne	1,800/2,600	06/11	*See case study Chap. 6*
Singapore	SingTel	1,800/2,600	12/11	
Singapore	StarHub	1,800/2,600	09/12	
Slovakia	Orange	1,800/2,600	–	*See case study Chap. 5*
Slovakia	Telefónica	1,800/2,600	–	
Slovakia	T-Mobile	1,800/2,600	–	
Slovenia	Telekom Slovenije	1,800	03/13	*See case study Chap. 5*
Slovenia	Telekom Austria	1,800	07/12	
South Africa	Cell C	Existing	–	*See case study Chap. 7*
South Africa	MTN	1,800	12/12	
South Africa	Vodacom	1,800	10/12	
South Africa	Telkom	1,800/2,300	04/13	
South Korea	KT Corp.	1,800	01/12	*See case study Chap. 6*
South Korea	LG Uplus	800/2,100	07/11	
South Korea	SK Telecom	800/1,800	07/11	
Spain	Orange	Multiple	06/13	*See case study Chap. 5*
Spain	Telefónica	Multiple	09/11[d]	
Spain	Vodafone	Multiple	09/11[d]	
Spain	TeliaSonera	1,800	07/13	
Sri Lanka	Etisalat	n/a	–	*See case study Chap. 6*
Sri Lanka	Mobitel	1,800	12/12	
Sri Lanka	Axiata	1,800	04/13	
St Kitts & Nevis	Digicel	–	–	Marketing HSPA + as '4G'
Sweden	Tele2	900/2,600	11/10	*See case study Chap. 5*
Sweden	Telenor	900/2,600	11/10	
Sweden	TeliaSonera	Multiple	12/9	
Sweden	Hi3G Access	2,600	12/11	

(continued)

Table 2.1 (continued)

Country	Operator	MHz Band[b]	Launch	Comment
Switzerland	Swisscom	Multiple	11/12	*See case study Chap. 5*
Switzerland	Sunrise	Multiple	06/13	
Switzerland	Matterhorn Mobile	Multiple	05/13	
Taiwan	Chunghwa telecom	700/2,600	2013	*See case study Chap. 6*
Taiwan	FarEasTone	698–806	2015	
Tajikistan	MegaFon	1,800	–	Trials in early 2011
Tajikistan	Babilon-Mobile	1,800/2,100	10/12	(See Telecom.paper 2012b)
Tanzania	Smile Comms	790–862	06/12[h]	*See case study Chap. 7*
Tanzania	Vodacom	n/a	–	
Thailand	True Move	2,100	05/13	*See case study Chap. 6*
Trinidad & Tob.	TSTT	n/a	–	*See case study Chap. 7*
Turkey	Turkcell	n/a	–	*See case study Chap. 7*
Turkey	Avea	1,800	–	
Turkey	Vodafone	n/a	–	
Turks & Caicos	Digicel	700	–	*See case study Chap. 7*
Turks & Caicos	Islandcom	700	–	
UAE	Du	1,800	07/12	*See case study Chap. 7*
UAE	Etisalat	1,800/2,600	12/11	
Uganda	Smile Comms	790–862	06/13	*See case study Chap. 7*
Uganda	MTN	2G	04/13	
UK	EE	Multiple	10/12	*See case study Chap. 4*
UK	Hutchison	1,800/800	2013	
UK	Vodafone	800/2,600	08/13	
UK	Telefónica	800	2013	
Ukraine	MTS	n/a	–	*See case study Chap. 7*
Ukraine	VimpelCom	GSM	–	
Uruguay	Ancel	AWS	12/11	*See case study Chap. 7*
USA	AT&T	698–806	09/11	*See case study Chap. 3*
USA	Verizon Wireless	700/AWS	12/10	
USA	Sprint Nextel	1,900/850	07/12	
USA	T-Mobile US	AWS	03/13	
US Virgin Isles	Sprint Nextel	1,900	2013	*See case study Chap. 7*
US Virgin Isles	AT&T	698–806	07/13	
Uzbekistan	MTS	2,600	07/10	*See case study Chap. 7*
Uzbekistan	TeliaSonera	2,600	08/10	

(continued)

Table 2.1 (continued)

Country	Operator	MHz Band[b]	Launch	Comment
Uzbekistan	VimpelCom	2,600	12/11	
Venezuela	Digitel	1,800	–	*See case study Chap. 7*
Venezuela	Móvilnet	n/a	–	
Venezuela	Telefónica	n/a	–	
Vietnam	–	n/a	–	*See case study Chap. 6*
Zambia	Zamtel	n/a	–	*See case study Chap. 7*
Zimbabwe	Econet	Existing	–	*See case study Chap. 7*

[a]Data that cannot be reliably confirmed have been excluded. Comments in italics refer to country as a whole, otherwise to the individual operator. Operators are identified wherever possible by the name of the main shareholder in order to reduce confusion and to allow for easy comparison of where the main international operators are rolling out LTE. To ensure comparability, the table includes only those operators with national mobile licences used for retail purposes while certain of the other companies listed below are also discussed in the case studies. MVNOs are excluded unless they are national operators with their own 2G networks. Among those excluded for one reason or another are Aquafon (controlled by MegaFon) which received a licence to operate a LTE network in the 800 MHz band in July 2013 (Abkhazia/Georgia); NBN Co (which launched TD-LTE in April 2012), Virgin Mobile (which launched as a MVNO in September 2012 over the Optus network), iiNet [which launched in October 2012 over the Optus network—see Telecom. paper (2012c)], Exetel (which launched as a MVNO over the Optus network in February 2013; Internode (which launched as a MVNO in July 2013 over the Optus network) (Australia); Scartel/ Yota (which launched in May 2012 but announced in June that it would terminate services) (Belarus); Clearwire (Belgium); CTBC Cellular and Sky Brasil (which launched TD-LTE in the 2.6 GHz band in December 2011), On (which launched TD-LTE in the 2.6 GHz band in March 2013) (Brazil); Eastlink [which launched in February 2013—see TeleGeography (2013c)], MTS Allstream [which launched in September 2012—see TeleGeography (2012h)], Sasktel (which launched in January 2013) and Wind Mobile (Canada); VelaTel/NGSN (China); Pirkanmaan (Finland); DiscoveryTel (Ghana); Augere, Tikona Digital (India); Regional Telecom [which launched in June 2013—see TeleGeography (2013f)] (Iraq); Japan Communications (MVNO which launched in March 2013) (Japan); InterDnestrCom (see footnote 3) (Moldova); Tele2 (which launched in May 2012), Ziggo4 (which launched in May 2012) (Netherlands); Hafslund Telekom (Norway); Cyfrowy Polsat (which launched as a MVNO in August 2011), Milmex (which launched in the 3.5 GHz band in August 2011) (Poland); Osnova, Rostelecom [which launched in June 2013—see TeleGeography (2013e)], Scartel/Yota [which launched a wholesale operation in April 2012 hosting MegaFon from April (TeleGeography 2012f) and VimpelCom from December (Telecom.paper 2012d)], Skytel and Vainakh (which launched in Chechnya in January 2013 only to run into objections from the regulator—see TeleGeography 2013b) (Russia); Canal Digital [which launched as a MVNO over parent Telenor's network in December 2012—see TeleGeography (2012m)] (Sweden); UK Broadband (which launched TD-LTE in the 3.5 GHz band in October 2012) (UK); Agri-Valley Broadband (and subsidiary miSpot which launched in March 2013), Alaska Communications [which launched in October 2012—see TeleGeography (2012i)], Appalachian Wireless [which launched in June 2013—see TeleGeography 2013g)], BendBroadband [which launched in May 2012—see TeleGeography (2012e)], Big River Broadband (which launched in May 2012); Bluegrass Cellular (which launched in November 2012); Cellcom (which launched in April 2012), C Spire Wireless [which launched in September 2012— see Cellular-news (2012a)], FreedomPop (MVNO on the Sprint Nextel network); Leap Wireless (which launched in December 2011), MetroPCS (which launched in the 1,900 MHz and AWS bands in September 2010), Mosaic Telecom (which launched in July 2011), Panhandle (which launched in March 2012), Penasco Valley Telecommunications, Peoples Telephone Cooperative (which launched in February 2012), Pioneer Cellular [which launched in April 2012—see

TeleGeography (2012d)], Shenandoah Telecommunications [which launched in November 2012—see TeleGeography (2012l)], Sprocket Wireless (which launched in November 2012), Strata Networks [which launched in November 2012—see TeleGeography (2012k)], Thumb Cellular (which launched in December 2012), Ting [which launched as a MVNO on the Sprint Nextel network in August 2012—see TeleGeography (2012g)], US Cellular (which launched in March 2012) and the NetAmerica Alliance [see TeleGeography (2012c)] (USA). In November 2012, Space Data Corp. and Lemko Corp. launched in Alaska what was claimed to be the world's first satellite-based LTE network (TeleGeography 2012j). In December 2011, Saima Telecom launched in Bishkek, Kyrgyzstan over what appears to be a fixed wireless network. On the TD-LTE front, Zoda Fones announced that it would be launching a network in Abuja, Nigeria in January 2012 (Telecom.paper 2012a)

[b]AWS varies according to region but normally includes spectrum in the 1,700 MHz band paired with spectrum in the 2,100 MHz (2.1 GHz) band. 2,600 MHz (2.6 GHz) signifies the 2,500–2,690 MHz band (but is often referred to as the 2,500 MHz band). A useful discussion of how this band developed from its origins as the IMT-2000 expansion band can be found in Shah (2010). The 700 MHz band is stated in full elsewhere in the table as 698–806 MHz (digital dividend band in the Americas)—but note that it is divided into several sub-bands (see Chaps. 3 and 8)—while 800 MHz represents the 790–862 MHz (digital dividend) band in Africa, Europe and the Middle East. It is hoped that there will be agreement to add a 700 MHz digital dividend band comprising 703–733 MHz paired with 758–788 MHz. The digital dividend band in Asia spans 470–960 MHz although many individual countries have selected the 700 MHz band as specified by the Asia-Pacific Telecommunity (APT) as 703–748 MHz paired with 758–803 MHz—see Global mobile Suppliers Association (2010); 4G Americas (2011: pp. 64–9) and the Mexico case study. The 850 MHz band was used for iDEN in the USA but is being switched to other technologies. The 900 MHz band combines an uplink of 880–915 MHz with a downlink of 925–960 MHz. The 1,800 MHz band combines an uplink of 1,710–1,785 MHz with a downlink of 1,805–1,880 MHz. The 1,900 MHz band combines an uplink of 1,850–1,910 MHz with a downlink of 1,930–1,990 MHz. The 2,100 MHz band combines an uplink of 1,920–1,985 MHz with a downlink of 2,110–2,170 MHz. The 2.3 GHz band spans 2,300–2,390 MHz. The 3.5 GHz band spans 3,400 MHz to 3,600 MHz. The 3.7 GHz band spans 3,600 MHz to 3,800 MHz. For further information specifically on GSM frequency bands see Wikipedia (2012)

[c]Awarded a 9 year licence, costing $2.7 million for 40 MHz paired. Technically the launch needed to be before 03/11 (MTS 2010) but did not take place until January 2012 (TeleGeography 2012a)

[d]These launches were for corporate clients rather than trials, but whether they should be considered as 'commercial' is debatable as discussed in the relevant case studies. The alternative dates for Vodafone are the Netherlands in February 2013, Spain in May 2013 and Greece in June 2013. KPN's launch in the Netherlands becomes February 2013

[e]As shown in the case study, SoftBank could be said to have launched two independent networks—the second in September 2012—as one was its own and the other obtained when it bought Willcom

[f]In late June 2013, Qatar Telecom officially changed its registered name to Ooredoo. However, given that Ooredoo is unlikely to become commonly used in the public domain for some time to come, the text uses the previous name to avoid confusion

[g]InterDnestrCom (Mobilink), which operates in the breakaway province of Transnistria, launched in the 800 MHz band in April 2012 (LteWorld 2012). The downlink only provided 10 Mbps

[h]It later stated that this had been a 'soft' launch with the 'commercial' launch taking place in May 2013

Source: Compiled by authors

not launched there; numerous roll-outs/trials such as those by Qatar Telecom, Telekom Austria (Bulgaria) and Telenor (Montenegro) listed as commercial launches; and TMN (Portugal) listed twice in March 2012.

2.2 Use of Multiple Bands

It is important to bear in mind that whereas 3G was largely allocated bandwidth not already in use for 2G—which in any event was often too congested to cope with the additional demands of data-rich downloads—the situation with respect to 4G is far more ambiguous. Certainly, the intention was always there to open up new spectrum bands, in particular the 2.6 GHz and digital dividend bands, but a widespread failure to make these available in good time meant that, with 3G provision operating independently, there was an opportunity to re-farm 2G bandwidth for 4G. In practice, this has so far primarily involved the 1,800 MHz band as shown in Table 2.1. As a consequence, the case studies often involve multiple spectrum bands any one, or indeed all, of which may be brought into play for LTE at some point.

2.3 Auction Methods

Historically, spectrum was mostly assigned without much regard to how demand for what is ultimately a finite band of frequencies useful for mobile communications was likely to develop. Hence, broadcasters and the military took possession of wide swaths of spectrum at zero or negligible cost and it became increasingly obvious that this initial set of assignments was not economically efficient. Market mechanisms became increasingly pervasive with the liberalisation of telecommunications commencing in the 1980s, and these were particularly associated with a switch to the use of auctions for assigning spectrum (Antonie and Colino 2011).

A brief comment about auction methods is accordingly needed at this point given that, unlike in the case of 3G spectrum, virtually all bandwidth suitable for 4G is being sold off—for an exhaustive discussion see DotEcon (2009: pp. 52–65). In essence, auctions can take three main forms as follows:
- Simultaneous multi-round ascending (SMRA)
- Combinatorial clock (CCA)
- Sealed bid

A SMRA can also come in a variant with augmented switching. The key point about a combinatorial clock auction is that bidders can make mutually exclusive package bids, and hence it is clearly suitable for occasions where spectrum in several bands is being sold simultaneously.

An idea of how these methods have been used can be seen in relation to the award of 800 MHz licences in Europe as follows:
- Germany 2010, Italy 2011, Portugal 2011, Spain 2011—SMRA as part of multi-band auction.

- Sweden 2011—SMRA in single band but with addition of augmented switching (see Aetha 2012: p. 22).
- Denmark 2012—combinatorial clock in single band.
- Switzerland 2012—combinatorial clock as part of multi-band auction.

As can be seen—see also the data on the 2.6 GHz auctions in Europe in Aetha (2012: p. 31)—there is far from unanimity as to which kind of method should be used, and it would serve little purpose to examine in detail the reasons why a particular method was adopted in any individual case—it may be noted that combinatorial clock auctions can be so complicated that even the bidders struggle to understand the rules as in the case of the Austrian auction in 2013 (see Chap. 5). The bottom line is that there is no clear relationship between the method used and the amount raised when expressed in dollars per MHz per head of population ($/MHz/pop) which is the standard valuation procedure for comparing results (with the potential added refinement of adjustments to take account of inflation over time and/or the length of the licence awarded).

It is not altogether clear why auctions have been much more prominent compared to beauty contests in the case of 4G compared to 3G. Given that auctions are expected to favour incumbents, regulators have sometimes considered it desirable to take steps to encourage new entry. The two methods in common usage—as shown in the case studies—are to cap the spectrum available to incumbents and/or to reserve certain wavelengths exclusively for new entrants. These methods have drawbacks—for example, incumbents may be prevented from acquiring spectrum that they value more highly than a potential new entrant or the reserved spectrum may prove not to be optimal for a potential new entrant (Cramton et al. 2011). One evident consequence is that incumbents have mopped up the great majority of the available spectrum (Blackman et al. 2013) although in fairness there was not much significant new entry as a result of 3G allocations despite the widespread use of beauty contests (Curwen and Whalley 2006). In any event, the trend is towards a reduction in the number of incumbents and/or infrastructure sharing, at least in developed countries, so it is unlikely that new entrants would have been rushing into markets with large established incumbents.

2.4 Coverage Obligations

Individual countries have taken widely different views as to the necessity to attach coverage obligations, expressed either in terms of population or geographic area, and whether some or all licensees should shoulder the obligations. For the most part, these obligations have been introduced in order to ensure that rural communities are not neglected when LTE is rolled out, but this desire to increase social benefits necessarily conflicts with the desire to use spectrum in the most efficient manner. Furthermore, it affects the value of the spectrum itself if sold at auction. There are those who argue that the initial sale should be unencumbered, with a subsequent competition related to rural provision, and it is certainly the case that licence obligations will make spectrum trading more complicated.

2.5   LTE Launches and Commitments

Before turning in the chapters that follow to a detailed examination of progress towards LTE provision in individual countries, it is useful to provide a summary of activity on a worldwide basis. It should be borne in mind that there are roughly 225 countries/islands in the world which strictly need to be considered on an individual basis, and that on average each has roughly four operators, so the list of operators in the table is unsurprisingly growing on a weekly basis. Whether it will ever reach 1,000 entries is a moot point given the delays in introducing 3G in many countries—the total number of countries where 4G is reasonably well-advanced in June 2013 is 140, comprising a little over 60 % of the total, but the proportion of operators that are actively involved is somewhat lower. The table covers known commitments to launch during 2013H2 but does not disclose where operators have merely reached the point of expressing an eventual interest in rolling out LTE.

Certain issues need to be addressed in relation to the identification of networks in Table 2.1. Many discrepancies exist between the original names recorded as network licensees and those currently holding those licences. This complicates the data collection process, and reflects the extensive merger and acquisition activity that has occurred throughout the mobile telecommunications industry as well as operators changing their names for a variety of reasons. Operators themselves can be the source of confusion by applying for licences using the names of existing or newly-created subsidiaries or via joint ventures/consortia.

It is by no means uncommon to find multiple names being used simultaneously in the public domain in relation to a single network. These names may refer, for example, to the majority owner of the network, to a minority owner whose name is nevertheless used because it is better-known than that of the other owners, to the historic name of the original network, to the brand name of the main service(s) or to a marketing name that it is hoped will enhance brand recognition which may involve another operator without an equity stake (for example, Vodafone).

Furthermore, for a variety of reasons, the official name of an operator may evolve over time without due recognition being taken of this by the media. Thus, for example, Celtel evolved into MTC Kuwait which evolved into Zain. However, matters may be more complicated. Thus, for example, a network in Nigeria started out as Econet Nigeria (Econet being the primary owner). This was rebranded in principle as Vodacom (another African operator) when it agreed to buy Econet, but a dispute with the original owners led Vodacom to pull out and to the use of Vmobile as the brand until its purchase by MTC led to the introduction of Celtel as the brand. When MTC became Zain, it used its own name as the brand. The most recent change has resulted from the purchase of Zain Africa by Bharti Airtel in 2011 which is now using its airtel brand, although in the media the network is more often identified as Bharti rather than as airtel.

Finally, although a name may appear to have been used consistently, this may disguise a somewhat different reality—for example, the current AT&T is a quite different organisation from that which existed under the same name 10 years ago. At another level, the brand name may have been kept on after a successful takeover—for example, Orange Switzerland is not owned by Orange but by Matterhorn Mobile.

These various factors explain why readers will struggle to marry up any two databases of LTE licences—even the authors, who keep a historical record of operator/brand names, are periodically confused by the nomenclature used in media reports. Obviously, therefore, some kind of systematic approach is needed here and in Table 2.1 the underlying principle is to cite wherever reasonable the controlling owner of the network. This has the clear advantage that it permits the reader quickly to ascertain the countries in which any given operator is involved in the launch of LTE. But because this is not the system used by the media—in fact, as indicated, there is no system used by the media, merely randomly expressed choices—the case studies often provide alternative names that readers may come across, although this has of necessity to be limited in scope in order to avoid undue complexity.

References

4G Americas. (2011, October). *The benefits of using LTE in digital dividend spectrum*. http://www.4gamericas.com

Aetha. (2012). *Spectrum value of 800MHz, 1800MHz and 2.6GHz: A DotEcon and Aetha Report for Ofcom*. London: DotEcon.

Antonie, G., & Colino, D. (2011). How to allocate spectrum rights efficiently. *Cudernos Económicos de ICE, 81*, 195–214.

Blackman, C., Forge, S., & Horvitz, R. (2013). Liberating Europe's radio spectrum through shared access. *info, 15*(2), 91–102.

Cellular-news. (2012a). *C Spire Wireless launches LTE services in Mississippi*. Accessed September 10, 2012, from http://www.cellular-news.com

Cellular-news. (2012b). *Ethiopia's capital city to get LTE coverage*. Accessed December 17, 2012, from http://www.cellular-news.com

Cramton, P., Kwerel, E., Rosston, G., & Skrzypacz, A. (2011). Using spectrum auctions to enhance competition in wireless services. *Journal of Law and Economics, 54*(4), 167–188.

Curwen, P., & Whalley, J. (2006). Third generation new entrants in the European mobile telecommunications industry. *Telecommunications Policy, 30*(10–11), 622–632.

DotEcon. (2009, December 21). Liberalisation of spectrum in the 900MHz and 1800MHz bands. ComReg Document Number: 09/99c.

Engineering news. (2013). *Ericsson lands Unitel Angola LTE contract*. Accessed January 12, 2013, from http://www.engineeringnews.co.za

Global mobile Suppliers Association. (2010). *Digital dividend update*. Accessed April 1, 2011, from http://www.gsacom.com

Global mobile Suppliers Association. (2012). *Evolution to LTE Report*. Accessed September 13, 2012, from http://www.gsacom.com

LteWorld. (2012). *Interdnestrcom*. Accessed July 9, 2012, from http://www.lteworld.org

MTS. (2010). *MTS allocated LTE frequencies in Armenia*. Accessed December 15, 2010, from http://www.mtsgsm.com

Shah, N. (2010). *2.6 GHz spectrum and the next generation mobile broadband networks*. Accessed December 21, 2010, from http://www.technowizz.wordpress.com

Telecom.paper. (2012a). *Zoda Fones picks Huawei to launch LTE TDD in Nigeria*. Accessed January 18, 2012, from http://www.telecompaper.com

Telecom.paper. (2012b). *Babilon-Mobile launches LTE services in Dushanbe.* Accessed October 9, 2012, from http://www.telecompaper.com

Telecom.paper. (2012c). *iiNet offers access to Optus LTE network.* Accessed October 10, 2012, from http://www.telecompaper.com

Telecom.paper. (2012d). *Beeline secures LTE access on Yota network.* Accessed December 10, 2012, from http://www.telecompaper.com

TeleGeography. (2012a). *Armenia's VivaCell-MTS switches on 4G/LTE network.* Accessed January 3, 2012, from http://www.telegeography.com

TeleGeography. (2012b). *Movicel launches LTE in Cabinda 45 days ahead of schedule.* Accessed April 16, 2012, from http://www.telegeography.com

TeleGeography. (2012c). *NetAmerica launches new programme to drive 700MHz LTE deployments.* Accessed April 30, 2012, from http://www.telegeography.com

TeleGeography. (2012d). *Pioneer Cellular launches rural LTE network in Oklahoma.* Accessed May 5, 2012, form http://www.telegeography.com

TeleGeography. (2012e). *Round the bend? Oregon telco claims LTE launch.* Accessed May 18, 2012, from http://www.telegeography.com

TeleGeography. (2012f). *Minister eyes 2013 launch for open-access LTE network.* Accessed May 21, 2012, from http://www.telegeography.com

TeleGeography. (2012g). *Ting becomes first US MVNO to offer LTE.* Accessed August 31, 2012, from http://www.telegeography.com

TeleGeography. (2012h). *MTS Allstream launches LTE network.* Accessed September 26, 2012, from http://www.telegeography.com

TeleGeography. (2012i). *Alaska Communications to launch LTE in Anchorage, Fairbanks and Juneau.* Accessed October 4, 2012, from http://www.telegeography.com

TeleGeography. (2012j). *US LTE round up.* Accessed October 30, 2012, from http://www.telegeography.com

TeleGeography. (2012k). *Strata Networks set to launch LTE in Uintah Basin.* Accessed November 8, 2012, from http://www.telegeography.com

TeleGeography. (2012l). *Sprint affiliate ShenTel launches LTE in 56 markets.* Accessed November 27, 2012, from http://www.telegeography.com

TeleGeography. (2012m). *Telenor's Canal Digital offers 4G broadband to pay-TV customers.* Accessed December 18, 2012, from http://www.telegeography.com

TeleGeography. (2013a). *Evatis launches 3G+ in Djibouti.* Accessed January 3, 2013, from http://www.telegeography.com

TeleGeography. (2013b). *Roskomnadzor objects to Vainakh Telecom LTE launch in Chechnya.* Accessed January 23, 2013, from http://www.telegeography.com

TeleGeography. (2013c). *Eastlink mobile service launched after five year wait.* Accessed February 15, 2013, from http://www.telegeography.com

TeleGeography. (2013d). *Tigo launches HSPA+ technology under the '4H+' banner.* Accessed May 23, 2013, from http://www.telegeography.com

TeleGeography. (2013e). *Rostelecom launches LTE in Sochi.* Accessed June 4, 2013, from http://www.telegeography.com

TeleGeography. (2013f). *Regional Telecom LTE network goes live in Kurdistan.* Accessed June 5, 2013, from http://www.telegeography.com

TeleGeography. (2013g). *Appalachian Wireless unveils LTE service.* Accessed June 12, 2013, from http://www.telegeography.com

Wikipedia. (2012). *GSM frequency bands.* Accessed December 12, 2012, from http://www.en.wikipedia.org/wiki/GSM_frequency_bands

The USA

3

3.1 Introduction

The discussion below is primarily concerned with the four nationwide operators—AT&T (branded as AT&T Mobility), Sprint Nextel, T-Mobile USA and Verizon Wireless. All other operators in the USA are local/regional in terms of their own network coverage although Leap Wireless, for example, signed a wholesale agreement with Sprint Nextel in 2012 which in principle enabled it to achieve nationwide coverage (but to negligible effect).

Verizon Wireless (hereafter Verizon) has access to 20 MHz of 700 MHz band spectrum (identified as Band 13 which comprises 746–756 MHz paired with 777–787 MHz). On 5 December 2010, it launched its '4G LTE' service in 38 markets, planning to increase this to 78 markets by mid-2011. However, this was only available to those owning computers using Windows and acquiring a LG Electronic dongle for $100 in conjunction with a 2-year contract. Surprisingly, the cost was less than that for 3G with 5 GB, for example, costing $50 a month (an over-use charge of $10 per GB per month was applicable), but Verizon was keen for customers to transfer onto the new, more efficient network. The typical speed on offer lay between 5 Mbps and 12 Mbps downstream and between 2 Mbps and 5 Mbps upstream. As of the end of 2010, a significant issue existed in respect of the time being taken for its cdma2000 1xEV-DO network subscribers to transfer onto the LTE network, but Verizon was still able to lay claim to half a million subscribers by the end of 2011Q1. It had coverage of half the population by end-August and 200 million people by end-December—it is continuing its roll out into rural areas in collaboration with rural telcos that are signed up to the LTE in Rural America programme (Global mobile Suppliers Association 2012; TeleGeography 2011b, 2012r), with the earliest programme launches achieved by Pioneer Cellular and Cellcom in April 2012 (TeleGeography 2012b, e). It is also launching LTE as a home broadband alternative branded as HomeFusion Broadband (Telecom.paper 2012c). The latest coverage statistic claims an improvement to 90 % in February 2013.

P. Curwen and J. Whalley, *Fourth Generation Mobile Communication,*
Management for Professionals, DOI 10.1007/978-3-319-02210-9_3,
© Springer International Publishing Switzerland 2013

In March 2012, it announced that during the rest of the year it would only be unveiling smartphones capable of running on the its LTE network in an attempt to increase the LTE subscriber base, and that it hoped to double the number of markets with coverage by the year-end, leaving Alaska as the only state without coverage. In August, it claimed to have coverage of 371 markets and 75 % of the population.

It may be noted that Verizon's launch lagged that of regional operator MetroPCS which, unlike Verizon, provided from the outset a handset in the form of the Samsung SCH-R900 (Craft) although this was in reality a souped-up feature-phone with limited capabilities—Verizon first marketed the HTC Thunderbolt in March 2011 and followed up with the Galaxy Tab 10.1 tablet in July. However, their restricted coverage means that MetroPCS and other regional operators will not be given further detailed consideration except insofar as they are involved in mergers or other arrangements with the four majors. Meanwhile, AT&T had altered its strategy which, in early 2010, had consisted of an upgrade to HSPA with a 7.2 Mbps downlink followed by a switch to LTE in 2011. In May 2011, it announced that it would continue to upgrade its HSPA network to 14.4 Mbps at a modest cost of $10 million—further upgrades would be more costly because they would necessitate the use of MIMO antenna—while testing LTE in Dallas and Baltimore. However, the purchase of Qualcomm's FLO TV spectrum (see below) indicated that 'true 4G' was now the ultimate objective, and various rumours indicated that the initial launch would take place in New York in mid-2011. The reality turned out to be a launch of several (non-handset) devices in August prior to the official launch of the network (using the 700 MHz Band 17 comprising 704–716 MHz paired with 734–746 MHz) in September in Atlanta, Chicago, Dallas, Houston and San Antonio (Gabriel 2011e). Two handsets, the Samsung Galaxy S II Skyrocket and the HTC Vivid, were made available in October, and in January 2012 AT&T claimed that its network covered 74 million people.

It was expected that a considerable time would elapse before it could expect to catch up with Verizon, although it is worth noting that even by February 2012 Verizon had only induced 5 % of its customers to switch to its LTE network despite a promotion offering twice as much data at half the price (Bensinger 2012b). However, according to a recent announcement in January 2013, Verizon's LTE network at that time covered 90 % of its legacy 2G/3G footprint with more than 50 % of all data travelling across its LTE network.

In contrast, T-Mobile USA faced a major dilemma. Initially, it stated that it intended to eschew LTE until 2013 while launching the 42 Mbps version of HSPA+, and it signed a roaming agreement with AT&T once the takeover bid by AT&T had been terminated (Telecom.paper 2011c). However, although this triggered the transfer of both cash and spectrum from AT&T to T-Mobile USA (Sahota 2012), the latter continued to lose contract subscribers, in part because it was the only national operator not supplying the iPhone, and it was forced to lay off 5 % of its workforce in March 2012. As a result, it lacked the resources to implement a national roll-out of LTE, yet its competitive position was set to continue to erode if it failed to do so.

It embarked on a plan to have 2,500 cell sites LTE-enabled by the end of July 2012 with a view to the commencement of commercial services in 2013. It is of particular interest that it is working with Nokia Siemens Networks to utilise LTE-Advanced for the roll–out. The plan was to use its AWS spectrum for the LTE roll-out while re-farming its existing 1,900 MHz spectrum for HSPA+. The first launch of the latter service took place in Las Vegas in September 2012 with Seattle, Washington D.C. and New York City set to follow on, while the official launch of LTE took place in March 2013 (TeleGeography 2013c).

However, plans are necessarily going to change now the merger between T-Mobile USA and MetroPCS has closed after an improved offer tabled on 10 April 2013 cleared the way for a successful conclusion (Wood 2013) which has created a publicly-traded entity to be known as T-Mobile US owned 74 % by T-Mobile. Having previously failed to merge with both Leap Wireless and Sprint Nextel, it was a case of third-time lucky for MetroPCS, and the partners are clearly motivated by the desire to overtake Sprint Nextel. MetroPCS uses cdma2000 technology for 2G and 3G and is a wholly pre-paid operator, neither of which is the case for T-Mobile USA. However, both operate in the PCS and AWS bands, so the merger plan involves shutting down the MetroPCS cdma2000 network and switching subscribers to the now enhanced T-Mobile USA network which will steadily convert to LTE while also pushing HSPA+ as a fall-back (Gryta 2012a). This will all take place over a 2-year period and will enable the post-merger entity to deploy a 20 MHz paired LTE channel in 90 % of the top 25 markets. In principle, Sprint Nextel could have made a counter-bid before the merger closed, but it is still possible that it will make a bid for the post-merger entity.

3.2 Satellite Provision

A unique feature of the USA is the plan to launch a wholesale service using mobile satellite spectrum. Harbinger Capital Partners set up LightSquared using licences owned by subsidiary SkyTerra, and partnered with Inmarsat in order to create the necessary contiguous bandwidth to supply terrestrial services.[1] LightSquared then joined up with rural access start-up OpenRange to provide services in areas neglected by the 'big three', although competition has subsequently appeared from two rival consortia comprising the NetAmerica Alliance and LocalLoop/Runcom (Gabriel 2011c), and Dish Network has spent almost $3 billion buying up 40 MHz of contiguous S-band spectrum (2,000–2,020 MHz plus 2,180–2,200 MHz)—the S-band is known in FCC parlance as AWS-4.

Furthermore, there have for some time been widespread concerns over the potential for LightSquared to interfere with GPS signals (Fitch Ratings 2011; Gabriel 2011d)—LightSquared hopes to use 1,545–1,555 MHz (upper 10 MHz

---

[1] See the series of articles on LightSquared by C. Gabriel at http://www.rethink-wireless.com, for example those on 21 July 2010, 6 September 2010 and 18 December 2010.

band) while GPS operates in the 1,559–1,610 MHz band. LightSquared has sought to downplay the issue while at the same time accusing the GPS industry of failing to comply with the Department of Defense's filtering standards—evidently the fact that adjacent spectrum was unoccupied provided the incentive to ignore compliance (Cellular-news 2012b). These concerns were first expressed by regulators in early 2011, but the FCC went ahead with approval for the non-satellite usage (Gabriel 2011a). In March 2011, LightSquared announced that it had 60 potential partners, among them Leap Wireless and Best Buy, but no single major client. It signed up InterGlobe Communications and Simplexity MVNO Services in August.

As things stood at the end of August 2011, the Department of Defense standard of September 2008 effectively granted GPS a 4 MHz guard band, LightSquared had offered a 23 MHz guard band and the GPS industry was asking for a 34 MHz guard band which would leave LightSquared with only 24 MHz of usable spectrum. In October, LightSquared proposed its third solution to the interference problem involving a wideband antenna developed by Petel, but the outcome remained in doubt (TeleGeography 2011c).

In early January 2012, Sprint Nextel gave LightSquared a final 30-day reprieve to obtain authorisation for its satellite service. This was a significant issue for Sprint Nextel given that LightSquared had agreed in July 2011 to pay $9 billion over an 11-year period for Sprint Nextel to manage the roll-out of its wholesale network, although the operator had other irons in the fire such as its arrangements with Clearwire (see below). However, it is possible that LightSquared's prospects have now been damaged irreversibly by the National Executive Committee for Space-Based Positioning, Navigation and Timing (PNT ExComm) which claimed that there could be interference with aircraft safety systems and that there was no solution to this problem (Cellular-news 2012a). The owners of LightSquared responded that the investigation was set up in such a way as to pre-determine the eventual findings, which induced Sprint Nextel to agree a further 6 weeks extension, but this may be to no avail even though the FCC agreed to a public consultation on the issue of whether GPS should have legal protection (Gabriel 2012b).

Indeed, in late February 2012, the FCC responded to a letter from the National Telecommunications and Information Administration (NTIA), which contained a summary of tests conducted so far, by announcing that it intended to revoke LightSquared's provisional permission to proceed (Telecom.paper 2012a). This, in turn, was expected to affect Sprint Nextel's agreement with LightSquared and it was duly scrapped by Sprint Nextel in late March (Cellular-news 2012d) with highly negative effects upon market perceptions of its financial health. Predictably, LightSquared filed an appeal, accusing the FCC of a 'bait and switch…of historic scale'. It continued to claim that testing had been conducted in respect of an entirely theoretical network that had some significantly different characteristics to its actual network (Cellular-news 2012f), but this was ignored.

What at the time appeared to be the last spin of the dice for LightSquared took the form of an attempt to switch some spectrum with that used for aircraft testing by the Department of Defense. Its financial situation began to look increasingly precarious given that it was obliged to repay large sums of interest on loans during the course of 2012, and in late February it announced that it would be laying off 45 % of its staff. Furthermore, the writs had begun to be lodged claiming that LightSquared's owners had dissipated billions on a fruitless quest to set up the sate/ @4erllite network (Bensinger 2012a). A voluntary bankruptcy was expected to be the next move, but even though not all developments were entirely negative, Chap. 11 bankruptcy rapidly became the inevitable outcome (TeleGeography 2012c)—a process which allows a company to continue in operation while it restructures its affairs. A debtor-in-possession loan in June 2012 means that the company can at least remain in business until November 2013 (Checkler 2012a)— that is, assuming the ongoing battles with lenders over various so-called insider loans do not cause everything to implode at an earlier date (Fitzgerald 2012b). The total losses accumulated during the year commencing May 2012 when it was declared bankrupt amounted to $661 million.

However, there is some evidence that LightSquared's plight has attracted a sympathetic response in some powerful quarters. In September 2012, a House of Representatives panel opened an investigation into whether the FCC followed its own rules when it initially authorised LightSquared in early 2011 (Checkler 2012b). For its part, LightSquared filed to modify its licence application to permit the use of the five megahertz of spectrum that was not held to cause interference problems together with a further five megahertz that it would share with the federal government while forgoing any further action in respect of the contested bandwidth (Checkler 2012c). In May 2013, the FCC authorised LightSquared to share a block of spectrum for 3 months with the National Oceanic and Atmospheric Administration in order to test whether they could co-exist.

However—and not for the first time as will become clear in what follows—Dish Network intervened to place on the table a so-called 'stalking horse' bid for LightSquared spectrum designed to set in motion an auction for LightSquared which would, in the particular circumstances of the bankruptcy, require court permission to be taken seriously (TeleGeography 2013e).

The situation faced by LightSquared can be compared with that of Dish Network itself, which acquired its S-band spectrum for $2.8 billion from bankrupt DBSD and TerreStar, subject to regulatory approval by the FCC and the resolution of Chap. 11 proceedings (Fitzgerald 2012a). The S-band is considered to be superior to the L-band used by LightSquared because it can be used twice—once terrestrially in what is known as an Auxiliary Terrestrial Component and once via a satellite. However, whereas LightSquared has circumvented the existing restriction on dual use via a waiver from the FCC (but run into interference problems), Dish Network has struggled to secure a satisfactory waiver—in March 2012 the FCC postponed a decision until the year-end, probably in order to avoid the kind of fall-out arising from its adverse decision about LightSquared (Gabriel 2012b, d).

Although Dish Network was expected eventually to get its waiver, there were concerns that it would need to address any issues potentially caused by the FCC's reassignment of AWS-4 spectrum (TeleGeography 2012n and see below). That said, its main problem remains a cash shortfall to fund a full terrestrial roll-out and a shortage of spectrum. It had begun talking in vague terms of a link of some kind with either Clearwire or Sprint, but many commentators argued that since the takeover of T-Mobile USA by AT&T had failed to come to fruition, Dish Network could be interested in merging its spectrum with that of T-Mobile USA—it had made an unsuccessful bid worth $4 billion for MetroPCS in August 2012 before T-Mobile USA interceded (Ramachandran 2012) and subsequently held informal discussions with T-Mobile USA about the possibility of a merger—or even offering to be bought by AT&T (Gabriel 2012a).

However, a link-up with AT&T seemed unlikely given the latter's petition to the FCC in February 2012 that Dish Network should have the same roll-out obligations as those imposed on LightSquared—coverage of at least 100 million PoPs (points of presence) within 33 months, 145 million within 45 months and 260 million within 69 months. Dish Network considered this to be a wholly unreasonable requirement on the grounds that, firstly, it intended to operate as a retailer rather than a wholesaler (which has ramifications for the prospects of 4G in the USA) and, secondly, that it intended to move straight to LTE-Advanced, a launch of which would not be feasible before 2015 (Gabriel 2012c). In May 2012, it announced that that it now expected to launch LTE-Advanced in 2016 with 60 % population coverage, but whether that would satisfy the FCC and result finally in permission to acquire the S-band spectrum remained in doubt (TeleGeography 2012d). As an intermediary step, Dish Network launched its upgraded dishNET satellite service, providing up to 10 Mbps downstream, on 1 October 2012 (Telecom.paper 2012e). However, it was then caught up in new rumours as Sprint Nextel hinted that it would be capable of—as against intent upon—hosting Dish Network on its terrestrial network.

The situation in November 2012 appeared to be that the FCC was prepared to sanction full terrestrial rights but only if onerous conditions were to be accepted by Dish. In effect, the FCC wanted Dish Network to disable 25 % of its uplink spectrum and impair a further 25 % in order to accommodate the potential future use of adjoining H Block spectrum by Sprint Nextel. As this 5 MHz block of spectrum was yet to be licensed, Dish understandably considered the proposal to be unreasonable and likely to cause yet further delays in its roll-out. The official FCC clearance finally came in mid-December with the expected conditions attached as well as a requirement that Dish Network build out at least 70 % of its network within 5 years. Dish chose not to respond at the time. The coverage condition was subsequently reported as 40 % of the population within its footprint within 4 years and 70 % within 7 years (TeleGeography 2012s).

As was widely predicted, the offer by Sprint Nextel to buy the outstanding shares in Clearwire at $2.97 apiece effectively forced Dish Network into making an unsolicited counter offer worth $3.30 per share or $4.85 billion for the entire shareholding. This was subject to Dish Network acquiring 25 % of Clearwire's spectrum for $2.2 billion in cash and included an agreement whereby Clearwire would build and operate the Dish Network network using some of Clearwire's spectrum and obtain new funding for Clearwire (FitzGerald 2013). In response, Clearwire noted that some of Dish Network's proposals might not be permitted under the terms of Clearwire's existing legal and contractual obligations. However, splitting Clearwire's spectrum between two rival operators could prove to be a blessing when the regulators cast their eyes on the various takeover proposals (Lennighan 2013a).

More recently, another, relatively small, satellite operator filed with the FCC to use its airwaves to provide mobile broadband. In this case, Globalstar intends, if licensed, to partner with an as yet unspecified company, taking advantage of any precedent established in respect of Dish Network. In order to overcome any objections about potential interference with GPS signals, Globalstar intends to use any bandwidth in the vicinity for a relatively weak uplink signal (Gryta 2012d).

3.3   HSPA+ Versus LTE in the USA

Some analysts have questioned whether any of the US operators should be interested in launching LTE. According to Aircom International, it would take time for LTE to achieve the (actual as against theoretical) speeds available via HSPA+, and although LTE would eventually catch up and would provide better spectral efficiency, and hence a cheaper service, this had to be set against the initial capital outlays for the two technologies. According to Aircom (Fitchard 2010), an upgrade of a 3G network to HSPA+ would cost $595 million whereas the construction of a nationwide LTE network would cost $1,780 million. This figure could be reduced by restricting coverage, but that could lead to complaints about poor service.

However, this argument was thrown into question in December 2010 when AT&T (provisionally) bought from Qualcomm (for $1,923 million) the extensive spectrum that it had won in the auctions for blocks in the Lower 700 MHz band—Auctions 44 and 49 in 2002 and 2003. Qualcomm had used this spectrum to launch FLO TV, a direct-to-handset mobile TV service, but had attracted so few customers that it had decided to close the service in March 2011 and sell on the spectrum at a substantial profit. The regulatory conditions were eventually satisfied in December 2011—the FCC had announced in August 2011 that the deal would be considered in conjunction with AT&T's ultimately unsuccessful take-over bid for T-Mobile USA. The deal enabled AT&T to roll out so-called 'true 4G' even though its other (paired) 700 MHz holdings are in a different bandwidth (AT&T 2010). It is worth adding that this event has brought to the fore the need for a means to use paired and unpaired spectrum in the same band for a single service—the UMTS licences in Europe often provided TDD spectrum which operators had mostly left unused, so this issue had been left on the backburner for far too long.

3.3.1   Is There a Spectrum Shortage in the USA?

The overall situation in the USA is clouded by the existence of local/regional operators and the role played by satellite operators, so it is apposite to ask just how much spectrum is available for 4G. According to a report published by Citi Investment Research & Analysis, a total of 192 MHz was being used in 2011Q3 out of a total availability of 538 MHz, with a further 300 MHz due for subsequent release (Davies 2011). The report claimed that 90 % of the spectrum in use was given over to 2G, 3G and HSPA+ and could not conveniently be re-farmed for 4G. It is also worth noting that clearing new spectrum bands may well cost more than can be recovered from auction revenues (Cellular-news 2012e). Further, even if the total available spectrum was to be allocated to 4G, the downlink available at busy times would only approximate to 5 Mbps which hardly meets the definition of 4G. Hence, if this report is taken at face value, the overall conclusion must be that while there is no overall shortage of spectrum, terrestrial operators will struggle to provide a high-speed 4G service throughout the USA unless they can rationalise their spectrum into efficient 20 MHz carriers, thus providing an opportunity for satellite providers to play a significant role in providing 4G services.

A somewhat different perspective is presented in a recent report by 4G Americas (Gabriel 2012e). This claims that, as of May 2012, the USA had less bandwidth per subscriber than other advanced economies—a claim based on NTIA data showing that there are 788,000 subscribers per MHz in the USA and that there is only 50 MHz of usable spectrum in the pipeline. For its part, the FCC takes the view that the government needs to re-designate some of its spectrum holdings for commercial use and share them with operators. In particular, the 1,755–1,850 MHz band has been identified as suitable by the NTIA (Gabriel 2012f). In April 2012, a bill was introduced in Congress (H.R. 4817) proposing that this band be paired with 2,155–2,180 MHz (designated AWS-3 by the FCC) and auctioned by February 2015 (Global mobile Suppliers Association 2012). In August 2012, the FCC granted temporary permission to T-Mobile USA to conduct LTE tests in the band. However, the latest development has come by way of an announcement by the FCC in March 2013 that it intends to auction 1,695–1,710 MHz paired with 1,755–1,780 MHz, possibly in September 2014.

One possibility is to open up spectrum bands that have so far largely been ignored for LTE. For example, AT&T is keen to utilise the 2.3 GHz band—which is popular in Asia—because of its spectrum holdings in that band (known as Wireless Communications Services (WCS) and originally licensed in 1997)—the other main holder is NextWave Wireless (see below). However, the problem is that the band is bisected by satellite radio transmissions, and LTE signals could potentially create interference. The FCC sanctioned the use of the band in principle in May 2010, but its mooted rules on avoiding interference proved objectionable to AT&T on the grounds that it left no opportunity to use the spectrum efficiently. As a result, AT&T joined forces with satellite radio operator Sirius XM to press for a solution which protects radio signals. It proposed that two 5 MHz bands of unpaired spectrum in proximity to that of Sirius be left as guard bands while releasing the two 5 MHz paired bands further away for LTE (Gabriel 2012g).

In October 2012, the FCC tabled its latest set of plans which set out to make a further 300 MHz of spectrum available by 2015. The plans largely concerned the AWS band, with a mooted initial auction of the so-called 'AWS-2 H Block', of particular interest to Sprint Nextel. However, there was inevitably a complication in that this would require a shift in the spectrum holdings of Dish Network (see discussion above), rendering a section incompatible with existing LTE standards (Gabriel 2012k).

One innovative idea to emerge from the FCC concerns so-called 'incentive auctions'. In effect, the objective is to provide an incentive for the holders of under-utilised spectrum to hand it over for re-farming in return for receiving part of the proceeds of any subsequent auction of the spectrum (Telecom.paper 2012d). The current plan is to finalise formal rules for the 600 MHz band by mid-2013 and to hold auctions during 2014. Initially, harking back to the 2009 National Broadband Plan which concluded that 120 MHz could be freed up using 6 MHz channels in the UHF TV spectrum band, the plan involves a two-stage auction. The first part will involve acquiring the spectrum from the broadcasters via a reverse auction in which at least two competing licensees must offer a minimum acceptable sale price and, after a repackaging exercise, the second part will involve the sale of the repackaged spectrum to mobile operators (Gabriel 2012j). Broadcasters will be able to retain their spectrum and will be provided with guard bands to prevent interference, and this has led commentators to conclude that the owners of the most valuable licences, holding between 40 MHz and 60 MHz in total, will refuse to sell—for further details see Standeford (2013a, b, c). A variant of the FCC plan has been tabled by T-Mobile (TeleGeography 2013b).

As of February 2012, the situation was not looking too promising. Basically, the Jumpstarting Opportunity with Broadband Spectrum (JOBS) Act of 2011 was—in typical American fashion—part of a massive measure (H.R. 3630) concerned with job creation and middle-class tax relief. It passed through the House of Representatives in December 2011 and moved on to the Senate, but the usual internecine strife between Democrats and Republicans means that whereas there is broad agreement on the central provisions of incentive auctions, there is total disagreement on the fine print. The main sticking points for Democrats were provisions barring the FCC from setting aside any of the reclaimed spectrum for unlicensed use and from applying network neutrality rules to any spectrum licensed via an incentive auction. The odds favoured the omission of any provisions for incentive auctions when the legislation finally reached a vote in the Senate.

There are also ongoing discussions between the Department of Defense and AT&T Mobility, Verizon Wireless and T-Mobile US (but not Sprint Nextel) on the issue of sharing the 95 MHz of spectrum in the 1,755–1,850 MHz band allocated to the Pentagon and other federal agencies (TeleGeography 2013a).

It is unlikely that there will ever be unanimity on the issue of spectrum shortage, although that was the overall view expressed in two 2012 reports by, respectively, The President's Council of Advisors on Science and Technology (PCAST) and Mobile Future (Standeford 2013d), where the former argued that the spectrum shortage is an illusion created by poor management of federal bandwidth.

Irrespective of whether a spectrum shortage really does exist, we still need to consider how/whether spectrum can be rationalised, taking into account that, in December 2011, Verizon Wireless paid, subject to regulatory approval, $3.6 billion to acquire the 122 AWS spectrum licences held by SpectrumCo (Telecom.paper 2011b) as well as 20 MHz of AWS spectrum from Cox Communications (which had sold out of SpectrumCo) for $315 million (see below). In September 2012, Verizon Wireless claimed that it now had enough bandwidth to last for 4–5 years. A further interesting development was the aborted (at literally the final hour) deal whereby Sprint Nextel would buy MetroPCS for $8 billion including debt taken over (Telecom.paper 2012b). Given Sprint Nextel's difficulties, discussed elsewhere, it comes as no great surprise that they sought to link up with a very large regional operator using cdma2000 technology, but for now it is not known why the deal fell through.

However, this was not the only M&A activity associated with Sprint Nextel at this time. On the one hand, as noted elsewhere, it was said to be monitoring the T-Mobile USA bid for MetroPCS with great interest, mulling over whether to rebid for MetroPCS or, more likely, to bid for a post-merger T-Mobile/MetroPCS entity. On the other hand, it was said to be a potential target for Japan's SoftBank (Wakabayashi 2012), and if SoftBank were indeed to bid for a majority stake then the situation would become somewhat complicated. In practice, in October 2012, SoftBank bid $20 billion for a 70 % stake, which was not expected to create regulatory problems. Sprint Nextel immediately responded by raising its stake in Clearwire (see below) so that, in effect, SoftBank was acquiring control of both companies (Osawa 2012).

It may finally be noted that when considering how much spectrum any given operator can hold in any given market as a result of a change of ownership, the FCC uses a 'spectrum screen'. The upsurge in activity noted above and below has led to calls for this screen to be reviewed. Needless to say, the big three and the small regional operators have opposing views about the desirability of spectrum caps.

3.4   The Controversy Over the 700 MHz Band

To help understand what this is about, it is necessary to summarise the position in respect of AT&T in the wake of the failed takeover of T-Mobile USA which forced it to cede valuable spectrum to its intended target. Needing to supplement its holdings, AT&T made an agreed takeover bid for NextWave Wireless in August 2012 (Gabriel 2012h). NextWave Wireless had severely over-stretched itself buying both spectrum in FCC auctions—it spent $4.7 billion in the 1996 auction of PCS licences without the wherewithal to actually pay up—and the likes of IPWireless to support its commercial exploitation of WiMAX. This had resulted in a stint in Chapt. 11 bankruptcy and further severe financial difficulties in 2011 when its remaining assets consisted of AWS, WCS and 2.5 GHz licences—it had previously sold its 700 MHz licences, mostly to Verizon Wireless. Its only way out was to be acquired, and AT&T obligingly offered $25 million for the company plus a contingency fee of $25 million and a further $550 million for its debt.

AT&T had also been buying up 700 MHz licences from rural cellcos (TeleGeography 2012g, o) and sought to make further AWS spectrum acquisitions from CenturyTel in September 2012 (TeleGeography 2012m), but as indicated above a national operator in the USA does not have a consistent amount of spectrum in any bandwidth on a nationwide basis and it is very difficult, if not impossible, to achieve this via piecemeal acquisitions. In theory, given AT&T's existing extensive holdings of WCS spectrum, combining these with the NextWave holdings will provide a platform for a nationwide roll-out—which it sought to expand via WCS purchases from Comcast and Horizon Wi-Com (TeleGeography 2012k)—but the need to resolve the interference issue discussed above remained an awkward issue. Needless to say, the FCC was petitioned to examine all of the acquisitions by AT&T as a whole rather than piecemeal, but it seemed doubtful that the FCC would want to raise further barriers to spectrum consolidation by AT&T and, indeed, the FCC duly approved the changes in the WCS band enabling AT&T to develop LTE using the 2,305–2,320 MHz and 2,345–2,360 MHz bands (TeleGeography 2012p).

In January 2013, AT&T also acquired spectrum in multiple bands from CDMA operator Atlantic Tele-Network (branded as Alltel) subject to regulatory review. However, it followed this with a much larger provisional deal involving 700 MHz B Block licences bought from Verizon Wireless for $1.9 billion covering 42 million people in 18 states. In addition to the cash payment, AT&T also handed over AWS licences in a number of markets. The spectrum, if authorised to be transferred in 2013H2, will be used by AT&T for LTE provision (Cellular-news 2013).

The most pertinent sections of the 700 MHz band are divided up as follows (Phone Factor 2008):

- Band 12: Lower A Block, Lower B Block, Lower C Block
- Band 13: Upper C Block
- Band 14: Upper D Block and Public Safety allocation
- Band 17: Lower B Block, Lower C Block

In addition to the paired spectrum in the Lower A, B and C Blocks there is unpaired spectrum in the Lower D and E Blocks. The Lower C and D Blocks were auctioned off in 2002 and 2003 (Auctions 44 and 49) whereas the Lower A, B and E Blocks were auctioned off in 2008 (Auction 73). Auction 73 also saw the attempted disposal of the Upper C and D Blocks which failed in respect of the D Block.

As noted above, Verizon Wireless dominates Band 13 while AT&T dominates Band 17. Band 12 is largely the preserve of regional operators including MetroPCS. This is a crucial distinction for equipment vendors since both Verizon Wireless and AT&T Mobility have the requisite economies of scale to justify the provision of equipment in the relevant band whereas that is not the case for the other operators. In 2012, a group of Tier 2 and Tier 3 operators, acting jointly as the 700 MHz Block A Good Faith Purchasers Alliance, filed with the FCC (rulemaking 11592) claiming that a lack of suitable equipment was preventing them from exploiting their spectrum, in part because Verizon Wireless and AT&T Mobility were pressurising vendors to concentrate on providing LTE equipment only in their Bands (FierceWireless 2013).

The Alliance requested that the FCC force vendors to produce equipment that works across all of the Bands. The nub of the defence was that Band 13 and Band 17 were sufficiently far apart as to ensure that equipment designed for one Band would not interfere with the other, and that interference would cause significant problems for multi-Band equipment. The two main vendors, Ericsson and Alcatel-Lucent, have so far largely kept their opinions to themselves although both are known to be developing LTE equipment for Band 14, and the FCC has yet to rule on the matter. It will need to take into consideration the fact that Verizon Wireless paid $2.5 billion for licences in Block A and hence appeared to have an incentive to push vendors to supplying equipment for that Block although, as noted below, it had promised to sell on this spectrum in return for permission to buy AWS licences from SpectrumCo so the issue may have little relevance. It will also need to have in mind its desire to foster nationwide LTE roaming and the fact that, although it strictly applies to AWS rather than the 700 MHz band, MetroPCS was not inhibited by the other, larger, AWS operators from being the first to launch.

3.5   WiMAX, Sprint Nextel and Clearwire

Does WiMAX have a future as a competitor for LTE? Until recently, it was arguable that only the USA was witnessing the roll-out of a network that might be seen in this light. When Sprint took over Nextel, it ended up with the great bulk of the spectrum in the 2.5 GHz band in the largest 100 markets in the USA. Given the existence of other users in the band, it appeared unlikely that Sprint Nextel could use it to roll out any kind of nationwide service, but the FCC imposed a condition for sanctioning the takeover of Nextel in the form of a requirement that the merged company offer services in the 2.5 GHz band to at least 15 million citizens by 2009 and to 15 million more by 2011. Technically, Sprint Nextel did not need to launch nationwide to achieve its FCC-mandated targets, but it appeared to be determined to achieve this although details were hazy until August 2006 when Sprint Nextel opted for mobile WiMAX using 802.16e-2005 technology.

In March 2007, AT&T sold its entire 2.5 GHz spectrum to Clearwire in order to satisfy one of the conditions attached to its purchase of BellSouth. This meant that, in effect, Clearwire and Sprint Nextel had the 2.5 GHz field to themselves although it is worth noting that whereas Clearwire now had access to extensive holdings of spectrum, this was highly fragmented and accordingly presented difficulties when attempting to pair up spectrum for FDD use. In July, Sprint Nextel and Clearwire announced that they now intended to build a joint nationwide WiMAX network branded as Xohm. Each would build out its own network and these would then be linked via a roaming agreement. By January 2008, Sprint Nextel was said to be desperately seeking partners in Xohm. In May came the announcement that the new partners would in fact be Comcast (which sold its shares in October 2012–it paid $550 million and got back $68 million), Time Warner Cable (which sold its stake in September 2012—it paid $550 million and got back $72 million), Bright House Networks, Intel (which wrote off its shares in 2012) and Google (which sold its stake in March 2012 at a loss of $443 million). They would get a combined 22 % stake in a new network branded as Clear, with Clearwire itself holding a further 27 % and Sprint Nextel the residual 51 %. The deal was sanctioned by the FCC in November 2008 and closed in December. Sprint Nextel chose to provide services over the Clear network acting as a MVNO branded as 'Sprint 4G'. The initial launch was in Baltimore at the end of September 2008.

At this time, mobile WiMAX had a first-mover advantage over LTE which was due on stream in 2010, and if it was a success in the USA it was expected to entice others to follow outside the USA where the future of WiMAX had been adversely affected by the decision of some major equipment vendors to concentrate upon LTE. However, it is fair to conclude, overall, that since mobile WiMAX has developed out of fixed WiMAX, it will remain of limited interest to mobile operators and that these will regard it as more equivalent to HSPA+ than to LTE.[2]

The primary evidence to support this conclusion is based on the progress shown by Clearwire which, in March 2010, appeared to be allowing for the possibility of at least a partial switch of technology in that it was one of the advocates of the opening up of the 2.6 GHz band for TD-LTE in the USA. Its target for WiMAX coverage in 2010 was 120 million people, of which 100 million were covered by the end of November. However, the rush to match the coverage of Verizon Wireless meant that it was running short of money and, if unable to raise new finance to fund its retail operations, it would have to behave as a wholesaler like LightSquared while leaving (primarily) 'Sprint 4G' to provide retail services. As it happens, Sprint Nextel had already expressed its displeasure at Clearwire's attempt to compete at the retail level given Sprint Nextel's majority stake in Clearwire, but suggestions of a full take-over were firmly denied. However, although Sprint Nextel appeared to be unwilling to provide any additional finance,[3] it claimed that Clearwire would be involved irrespective of the direction taken by its 4G strategy (Gabriel 2011b) and duly signed a new long-term agreement in April 2011 which specified wholesale prices and guaranteed minimum payments to Clearwire (Telecom.paper 2011a).

Subsequently, Clearwire announced that it would not be selling off any of its spectrum. In June 2011, Sprint Nextel reduced its voting rights in Clearwire to 49.9 % to allay shareholders' fears that it would be liable for debts incurred by Clearwire if the latter was deemed to be a subsidiary company (Cellular-news 2011b), and shortly thereafter signed a 15-year deal with LightSquared jointly to develop, deploy and operate its network (TeleGeography 2011a) although the questions over interference noted above were still to be resolved.[4]

---

[2] According to Clearwire (see 'The Clearwire Story' at http://www.clearwire.com) it is 'the pioneer in 4G and operator of the first 4G network in the country'. It goes on to refer to the proposed TD-LTE network as 4G. The operators, including Sprint Nextel and (most recently) EarthLink, that use it as a wholesaler also describe their services as 4G. However, the promised downlink is only 3–6 Mbps. It is also of interest that Digicel launched WiMAX in Jamaica using the brand 'Digicel 4G Broadband' (see Chap. 7).

[3] For an early assessment of Clearwire's prospects see, for example, Cellular-news (2011a). It may be noted that Clearwire was also awarded a licence in Puerto Rico in October 2009 covering 76.5 MHz of unpaired spectrum (as was James McCotter in Guam and the N. Mariana Islands).

[4] It is of interest that LightSquared began signing wholesale agreements despite its ongoing difficulties.

In August 2011, Clearwire confirmed that it would be overlaying its WiMAX network with LTE-Advanced-ready technology in the main urban areas provided it could secure the necessary funding. It reported that it had 7.7 million subscribers (of whom 1.3 million were retail and 6.4 million were wholesale) up from 1.7 million 1 year previously, and predicted 10 million subscribers by the year-end via an expansion of its wholesale operations. Its coverage encompassed 132 million people. At the time, Fitch Ratings provided an analysis of Clearwire, LightSquared and Dish Network (Fitch Ratings 2011) in which it described Clearwire's strategy as 'clear as mud' and opined that LightSquared 'remains grounded'. However, some light was shed by the announcement that Cox Communications and Cablevision System were considering the acquisition of a stake in Sprint Nextel which would use the cash to buy out the other shareholders in Clearwire.

In October, Sprint Nextel announced that it would not be selling any devices using Clearwire's WiMAX technology after 2012, although it would continue to support such devices as were already in circulation. Rather, it would be speeding up the roll-out of Network Vision which would combine its various technologies using the 1,900 MHz band, aiming for a soft launch in 2012Q2—achieved in April (TeleGeography 2012a). This was followed by a commercial launch—achieved in 15 cities including Atlanta, Dallas, Houston, Kansas City and San Antonio in July 2012 (TeleGeography 2012f, j)—and nationwide coverage by 2013 (by which point it expected to have re-farmed the 850 MHz spectrum previously used by its terminated iDEN service). This was well behind Verizon which claimed to have coverage for 200 million citizens in January 2012 and to be available in 304 markets covering two-thirds of the population in June 2012, rising to 400 markets in October. This had an ongoing negative effect upon Sprint Nextel's credit rating (Tadena 2011)—the fact that it was going it alone in the 1,900 MHz band meant at the very least that vendors would not be rushing to provide compatible devices—and elicited the common response that it appeared to have relegated itself to the status of permanent also-ran behind AT&T and Verizon given its relative paucity of spectrum (Davies 2011) even if Clearwire's somewhat fragmented TDD spectrum was not necessarily the answer to all of its prayers (Gabriel 2011f). For its part, Clearwire played down the significance of this development, pointing out that it remained the largest wholesaler in the USA and that Sprint Nextel would not be alone in needing to utilise Clearwire's spectrum at some point.[5]

There is clearly some truth in this claim since Sprint Nextel has begun to back-track. In early December, it agreed, subject to certain conditions, to pay Clearwire $926 million in return for unlimited 4G WiMAX retail services during 2012 and 2013. Commencing in 2014, usage-based pricing would be introduced. In addition, Sprint Nextel agreed to pay $350 million, via a series of pre-payments over a 2-year period, in return for LTE capacity provided that Clearwire achieved specified

---

[5] Clearwire was expected to be the main beneficiary of LightSquared's problems. In February 2012, it announced that it had reached 10.4 million customers at end-2011 and that calendar 2011 had seen revenues rise to record levels. However, as noted, this has not delivered financial stability.

roll-out targets by June 2013. The companies would jointly select where to locate equipment with an emphasis upon serving Sprint Nextel's hotspots. Finally, Sprint Nextel agreed to provide additional equity finance in the event of a further share issue (Cellular-news 2011c). This was needed because, in early December, the owners of SpectrumCo, a group of cable operators including Clearwire investors Time Warner and Comcast, sold their AWS licences to Verizon Wireless for $3.6 billion, subject to regulatory approval (Bensinger 2012c)—AT&T could not bid because of its ongoing merger negotiations with T-Mobile USA. This meant that they would cease to re-sell Clearwire's 4G services in March 2012. The deal received conditional approval from the Justice Department and FCC in August 2012 (TeleGeography 2012i), and Verizon agreed to sell on spectrum in the 700 MHz Lower A and B blocks to T-Mobile and/or other interested parties (TeleGeography 2012h).

Clearwire duly raised $715 million in mid-December, of which almost half was provided by Sprint Nextel (Cellular-news 2011d). In February 2012, it announced that it would complete the first phase of its roll-out, involving 5,000 cell-sites, by June 2013, and that a further 3,000 would be built in urban areas once these were operational, while in March it was able to reveal a 5-year contract to provide 4G capacity to major regional operator Leap Wireless (Stynes 2012). At the time, the plan was for Clearwire to continue to retail its own services while acting primarily as a wholesaler.

The ownership of Clearwire is currently in something of a state of flux given that, in March 2012, Google sold all of the shares that it had bought 4 years previously for $500 million, taking a heavy loss in the process, while Intel Corp. went so far as to write down the entire value of its 7.3 % voting stake. However, Sprint Nextel, with its near-half stake remained firmly in control with its right to nominate 7 of the 13 board members, although technically 10 votes in favour are needed for major decisions.

In August 2012, the Clearwire share price fell below $1, equivalent to a near-50 % fall during the year so far, in the process forcing Sprint Nextel to take an impairment charge. At 11 million, subscriber numbers appeared to be healthy, although the great majority were wholesale customers provided by Sprint Nextel. One interesting rumour was to the effect that Dish Network might be investing $400 million—it announced the investment in August 2012 but did not specify the target (Gabriel 2012i).

As noted above, the SoftBank takeover bid for Sprint Nextel (allegedly) caused the latter to raise its voting rights in Clearwire from 49.8 % to a small majority (without affecting its majority equity stake). This was something of a volte face given that the voting rights had previously been reduced to a minority in order to avoid Sprint Nextel taking Clearwire's debts onto its books (Cellular-news 2012g), but it did mean that SoftBank would gain control of two companies, each with significant spectrum holdings. It should be noted that the SoftBank bid, if success-ful, could affect Clearwire insofar that SoftBank is rolling out TD-LTE in Japan (Gryta 2012b). Certainly, Clearwire is now intent in slowing down the roll-out of its network in order to conserve cash and align it better with Sprint Nextel's own roll-out schedule which in October 2012 was running a quarter behind its year-end goal of 12,000 sites (Gryta 2012c). In November, Clearwire announced that it would be deploying only 2,000 of the 5,000 cell-sites originally pencilled in for March 2013 (TeleGeography 2012q).

In December, it was alleged that Sprint Nextel would proceed with a plan to acquire full control of Clearwire, causing the latter's share price to rise to $2.68. The implications for the takeover by SoftBank were not immediately clear, but in the event SoftBank expressed itself to be happy with a bid of $2.97 a share which was accepted by Comcast and Intel but, inevitably, other shareholders set out to contest the bid in the courts—a majority of non-Sprint Nextel shareholders were anyway required to support the bid if it was to be accepted. But as noted previously, all of these plans were thrown into disarray when Dish Network made an unsolicited bid for the whole of Clearwire in January 2013. Clearly, given its existing stake in Clearwire, Sprint Nextel could prevent a takeover by Dish Network and it could mount a higher counter-offer, but Dish was known to be litigious—it also asked the FCC to delay its review process which technically must be completed within 180 days commencing at the beginning of December—and hence the proposed links between SoftBank, Sprint Nextel and Clearwire would have to be adjusted in some as yet unspecified manner (FitzGerald 2013).

There were further developments in February 2013 when Clearwire announced that it needed additional funding, and hence that it intended to draw upon funds made available by Sprint Nextel as part of its attempt to take full control of Clearwire (Carew 2013). Sprint Nextel had offered to buy $800 million of bonds from Clearwire by way of 10 monthly instalments, a facility that Clearwire had not taken advantage of during January and February because Dish had stated that its own offer was contingent upon Clearwire not drawing upon the facility (Gryta 2012a). However, Clearwire did draw down the instalments commencing in March without eliciting any response from Dish Network.

In the event, it was Sprint Nextel that blinked first when presented with the likelihood that Clearwire's shareholders would reject its initial bid. In late May it raised its offer for the outstanding Clearwire shares to $3.40 each, thereby valuing the whole of Clearwire at $10.7 billion. It declared this to be its 'best and final offer' (Lunden 2013). Within days, Dish Network responded by raising its own bid to $4.40 a share in cash contingent upon acceptance by at least half of the minority shareholders (equivalent to one-quarter of the voting shares) and the offer of at least three seats on the board (Yu 2013). This raised the possibility that it would end up as a minority shareholder in a company controlled by Sprint Nextel.

Further complications arose in April when Dish Network made an (allegedly informal) unsolicited bid for 68 % of Sprint Nextel worth $25.5 billion—equivalent to $7 a share of which $4.76 consisted of cash, considerably more than the prior offer by SoftBank (Telecom.paper 2013a). SoftBank, in response, refused to raise its own offer, claiming that it expected it to be accepted on 1 July without further amendment. However, it soon changed its mind when shareholders reacted negatively, adding additional cash to its offer so as to raise it to $21.6 billion. Curiously, this meant that it would now end up with 78 % of Sprint Nextel, thereby reducing the value of the entire company from \$28.9 billion to \$27.8 billion (Gabriel 2013a).

Analysts were divided about the relative merits of the two offers for Clearwire. The main benefit of the Dish Network proposal was that it would create a portfolio of 230 MHz of spectrum across a range of bandwidths. This would enable it to provide a quad-play offer in competition with AT&T, Verizon and Comcast. However, there would be a need to invest heavily over a period of years to create the requisite network of cell sites, especially in urban areas, and to integrate the two companies (Lennighan 2013b). Given that Fitch had awarded Dish Network a BB-debt rating in 2012, indicating little confidence in its existing prospects, taking on further debt to finance the takeover was not seen by many as a positive move even though it did already have \$9.5 billion in cash and marketable securities at its disposal (Cellular-news 2011c). A factor worthy of further note was that the SoftBank proposal meant that it would basically become the controlling share-holder in Sprint Nextel in its existing form, whereas Dish Network and Sprint Nextel would have to undergo a problematic merger process.

As ever, the matter now moved into the courts, with Sprint Nextel seeking to block the Dish Network bid in mid-June. At first glance, the lawsuit appeared to have little merit other than as an attempt to discourage Clearwire shareholders from supporting the Dish Network offer. Nevertheless, it now seemed likely that, given its refusal to table a higher offer, Dish Network had given up on its pursuit of Sprint Nextel, preferring to devote its resources to acquiring Clearwire (Gabriel 2013b). If so, it was about to suffer probable rejection because Sprint Nextel raised its offer to \$5 a share, valuing the whole of Clearwire at \$14 billion, nearly three times its value in October 2012 before the bidding war began. In response, 9 % of Clearwire voting shares were irrevocably pledged to Sprint Nextel meaning that, as board members had switched their support from Dish Network to Sprint Nextel, the latter now had pledges for half of the independent voting shares (Telecom.paper 2013b).

Sprint Nextel also has alternative plans for LTE. In particular, it intends to market it via two existing pre-paid brands, Virgin Mobile USA and Boost Mobile, in an attempt to keep down the costs to customers—neither AT&T Mobility nor Verizon Wireless currently provide a pre-paid LTE option. In February 2013, Virgin Mobile offered for sale the Samsung Galaxy Victory 4G Android in con-junction with a 'No Contract Beyond Talk' unlimited data and messaging plan whereas Boost offered the HTC One SV Android and the Boost Force Android. For the time being, however, Sprint Nextel is struggling to roll out Network Vision, allegedly because of delays in vendor execution (TeleGeography 2013d).

The future prospects for WiMAX in the USA have also suffered due to a number of bankruptcies. For example, OpenRange filed for bankruptcy in October 2011 and Rocky Mountain Broadband in September 2012 (TeleGeography 2012l).[6]

---

[6] One further issue looks to be increasingly relevant. In February 2012, it was announced that public safety organisations in the USA would be allocated the previously unsold D Block spectrum in the 700 MHz band and that finance would be arranged to support the roll-out of LTE by these

Conclusions

As is evident from the above discussion, the situation in respect of high-speed data transfers in the USA is unusually complicated. It also has unique characteristics because of both the satellite issue and the major restructuring that is going on at the present time with unpredictable consequences.

As of mid-June 2013, the position was as follows.

- T-Mobile USA was merged with MetroPCS to form T-Mobile US
- Dish Network had offered to buy Clearwire
- Sprint Nextel had offered to buy Clearwire minorities
- Dish Network had offered to buy a majority of Sprint Nextel
- SoftBank had offered to buy a majority of Sprint Nextel
- Dish Network had made a 'stalking horse' offer to buy LightSquared spectrum

The total amount being committed came to comfortably in excess of $50 billion, but whereas that might indicate that there huge synergies there for the taking via any number of possible combinations between the protagonists—AT&T is also a potential player in the game of M&A at some point—it might also be taken as a sign that there is considerable uncertainty as to what might be termed a sustainable longer-term equilibrium structure for spectrum holdings in the USA.

It is self-evidently the case that M&A activity will proceed in stages, with the outcome of one deal itself affecting the outcome of those deals yet to be brought to fruition. It is probable that new deals will be engineered. It is also self-evident that regulators want to play an important role in the unfolding drama, a role that is itself unpredictable. Towards the end of June the first sign of matters moving toward some kind of resolution came in the form of public recognition by Dish Network that it was no longer an active bidder for Sprint Nextel. Sprint Nextel shareholders overwhelmingly voted in favour of the SoftBank offer and the FCC gave its approval in early July. The takeover was formally completed on 10 July. Interestingly, the SoftBank CEO stated that his main ambition was to speed up the Sprint Nextel network.

Dish Network also withdrew its offer for Clearwire (Telecom.paper 2013c), in effect admitting that the Softbank/Sprint Nextel/Clearwire link-up was a done deal. The FCC approved the Sprint Nextel/Clearwire deal which was completed on 8 July. As a fallback, Dish Network applied to the FCC to waive the roll-out obligations attached to the 700 MHz Block E spectrum that it had won in 2008 (TeleGeography 2013f). These were supposed to have been met by mid-June, but Dish Network argued that a combination of interference problems and lack of suitable equipment had forestalled any attempt to use the spectrum. However, this hardly represented an alternative strategy to the acquisition of Sprint Nextel and Clearwire, nor did the spectrum match any held by LightSquared. Hence, it

organisations. It is probable that most similar organisations around the world will now adopt LTE (Cellular-news 2012c).

was felt in some quarters that Dish Network would now turn its attention to some kind of link-up with one or more of T-Mobile US, Leap Wireless, nTelos Wireless and US Cellular.

However, Leap Wireless was promptly removed from the equation when AT&T made a takeover bid worth $1.2 billion in cash plus taking on $2.8 billion of debt (Telecom.paper 2013d). The deal included spectrum in the PCS and AWS bands covering 137 million people which would complement AT&T's existing holdings. The deal is expected to close in early 2014.

On a more positive note, Dish Network is set to gain from draft rules published at the end of June by the FCC in respect of the 1,900 MHz H Block (1,915–1,920 MHz paired with 1,995–2,000 MHz) (TeleGeography 2013g). This is hopefully to be auctioned (via Auction 96) no later than 2014Q1 and there will be no interference issues with Dish Network's S-band spectrum.

Ultimately, therefore, who will end up providing LTE and LTE-Advanced, to whom and by what methods is yet to be fully resolved in the very country that is one of the world leaders in launching new mobile technologies and which has certainly amassed far more LTE subscribers to date than any other country. The only fairly unequivocal statement that can be made is that Verizon Wireless is making most of the running in relation to LTE. In late June, it announced that its LTE network was substantially complete after reaching 500 markets, and that its thoughts were now turning to re-farming both its PCS spectrum and—in the face of customers increasingly switching from 3G to 4G—its AWS spectrum (TeleGeography 2013h). VoLTE would follow in 2014 and an all-IP future beckoned. Total subscriptions in July 2013 amounted to 35.7 million.

References

AT&T. (2010). *AT&T agrees to acquire wireless spectrum from Qualcomm.* Accessed December 20, 2010, from http://www.att.com

Bensinger, G. (2012a). *Harbinger investors sue Falcone, Harbinger fund over LightSquared.* Accessed February 16, 2012, from http://www.totaltele.com

Bensinger, G. (2012b). *Verizon Wireless still largely 3G a year after LTE debut.* Accessed February 28, 2012, from http://www.totaltele.com

Bensinger, G. (2012c). *Verizon, cable companies defend spectrum sale.* Accessed March 6, 2012, from http://www.totaltele.com

Carew, S. (2013). *Clearwire says needs Sprint funding to last to year-end.* Accessed February 12, 2013, from http://www.reuters.com

Cellular-news. (2011a). *Clearwire losses widen – may delay radio spectrum sale pending larger alliance.* Accessed February 19, 2011, from http://www.cellular-news.com

Cellular-news. (2011b). *Sprint Nextel reduces Clearwire voting rights to below 50%.* Accessed June 15, 2011, from http://www.cellular-news.com

Cellular-news. (2011c). *Clearwire gets financial lifeline from Sprint Nextel.* Accessed December 2, 2011, from http://www.cellular-news.com

Cellular-news. (2011d). *Clearwire raises $715.5 million to fund network upgrades and expansion.* Accessed December 14, 2011, from http://www.cellular-news.com

Cellular-news. (2012a). *Blow to LightSquared as government agency says GPS interference cannot be solved.* Accessed January 16, 2012, from http://www.cellular-news.com

Cellular-news. (2012b). *LightSquared ramps up pressure on GPS receiver manufacturers.* Accessed February 8, 2012, from http://www.cellular-news.com

Cellular-news. (2012c). *D Block allocation in the USA will open the door to mass private LTE.* Accessed February 21, 2012, from http://www.cellular-news.com

Cellular-news. (2012d). *Sprint Nextel scraps $9 billion network sharing deal with LightSquared.* Accessed March 20, 2012, from http://www.cellular-news.com

Cellular-news. (2012e). *US government could release 95Mhz of radio spectrum for mobile broadband services.* Accessed March 28, 2012, from http://www.cellular-news.com

Cellular-news. (2012f). *LightSquared responds to regulators' plans to block network launch.* Accessed April 2, 2012, from http://www.cellular-news.com

Cellular-news. (2012g). *Sprint buys back control of Clearwire.* Accessed October 19, 2012, from http://www.cellular-news.com

Cellular-news. (2013). *AT&T pays $1.9 billion for Verizon Wireless radio spectrum.* Accessed January 28, 2013, from http://www.cellular-news.com

Checkler, J. (2012a). *LightSquared lenders to give company $30m bankruptcy loan.* Accessed June 22, 2012, from http://www.telecomseurope.net

Checkler, J. (2012b). *US House panel looks at FCC's 2011 LightSquared approval.* Accessed September 24, 2012, from http://www.telecomseurope.net

Checkler, J. (2012c). *LightSquared proposes sharing wireless network with US government.* Accessed October 1, 2012, from http://www.telecomseurope.net

Davies, S. (2011). *There's no radio spectrum shortage in the USA – report.* Accessed October 3, 2011, from http://www.cellular-news.com

FierceWireless. (2013). *Will the FCC require all 700 MHz LTE equipment to interoperate?* Accessed January 31, 2013, from http://www.fiercewireless.com

Fitch Ratings. (2011). *The road to 4G.* Accessed August 3, 2011, from http://www.fitchratings.com

Fitchard, K. (2010). *Aircom: U.S. operators could save $1.2B by moving to HSPA+ rather than LTE.* Accessed May 28, 2010, from http://www.printthis.clickability.com

Fitzgerald, P. (2012a). *TerreStar wins approval of Dish-backed bankruptcy plan.* Accessed February 15, 2012, from http://www.totaltele.com

Fitzgerald, P. (2012b). *Falcone, Harbinger fire back at LightSquared lenders' probe.* Accessed August 9, 2012, from http://www.totaltele.com

FitzGerald, D. (2013). *Dish Network can still win, even if Clearwire bid fails.* Accessed January 10, 2013, from http://www.totaltele.com

Gabriel, C. (2011a). *LightSquared gets green light to launch network.* Accessed January 27, 2011, from http://www.rethink-wireless.com

Gabriel, C. (2011b). *Sprint says Clearwire is part of any 4G path it follows.* Accessed March 11, 2011, from http://www.rethink-wireless.com

Gabriel, C. (2011c). *Trio of alliances focus on rural wireless broadband.* Accessed March 27, 2011, from http://www.rethink-wireless.com

Gabriel, C. (2011d). *Two agencies warn of GPS interference from LightSquared.* Accessed June 14, 2011, from http://www.rethink-wireless.com

Gabriel, C. (2011e). *AT&T to launch LTE devices without a network.* Accessed August 17, 2011, from http://www.rethink-wireless.com

Gabriel, C. (2011f). *Sprint's 4G strategy condemns it to spectrum famine.* Accessed October 11, 2011, from http://www.rethink-wireless.com

Gabriel, C. (2012a). *Dish can play all sides in US 4G fall-out.* Accessed January 17, 2012, from http://www.rethink-wireless.com

Gabriel, C. (2012b). *LightSquared gets reprieve, AT&T argues over Dish.* Accessed January 31, 2012, from http://www.rethink-wireless.com

Gabriel, C. (2012c). *Dish needs more time to build LTE-Advanced network.* Accessed February 6, 2012, from http://www.rethink-wireless.com

Gabriel, C. (2012d). *Dish fails to win waiver from FCC*. Accessed March 5, 2012, from http://www.rethink-wireless.com

Gabriel, C. (2012e). *US really is running out of spectrum, says trade body*. Accessed May 18, 2012, from http://www.rethink-wireless.com

Gabriel, C. (2012f). *FCC chief calls for sharing of federal spectrum*. Accessed May 24, 2012, from http://www.rethink-wireless.com

Gabriel, C. (2012g). *AT&T proposes rules to make WCS band usable for LTE*. Accessed June 20, 2012, from http://www.rethink-wireless.com

Gabriel, C. (2012h). *AT&T to buy NextWave for its 4G spectrum*. Accessed August 2, 2012, from http://www.rethink-wireless.com

Gabriel, C. (2012i). *Dish's mystery partner may be Clearwire*. Accessed August 13, 2012, from http://www.rethink-wireless.com

Gabriel, C. (2012j). *FCC aims to fast track TV spectrum auction*. Accessed September 6, 2012, from http://www.rethink-wireless.com

Gabriel, C. (2012k). *FCC plans to add 300 MHz of spectrum by 2015*. Accessed October 8, 2012, from http://www.rethink-wireless.com

Gabriel, C. (2013a). *SoftBank lures Sprint investors with more cash*. Accessed June 11, 2013, from http://www.rethink-wireless.com

Gabriel, C. (2013b). *Dish abandons Sprint bid to focus on Clearwire*. Accessed June 19, 2013, from http://www.rethink-wireless.com

Global mobile Suppliers Association. (2012). *Evolution to LTE report*. Accessed September 13, 2012, from http://www.gsacom.com

Gryta, T. (2012a). *Spectrum needs at heart of any T-Mobile/MetroPCS deal*. Accessed October 3, 2012, from http://www.totaltele.com

Gryta, T. (2012b). *Clearwire shares jump as Sprint faces possible buyout bid*. Accessed October 11, 2012, from http://www.totaltele.com

Gryta, T. (2012c). *Clearwire clarifies Sprint relationship as loss widens*. Accessed October 26, 2012, from http://www.totaltele.com

Gryta, T. (2012d). *Globalstar to use spectrum for mobile broadband*. Accessed November 14, 2012, from http://www.totaltele.com

Lennighan, M. (2013a). *Dish Network's high-profile chairman is not just playing games*. Accessed January 11, 2013, from http://www.totaltele.com

Lennighan, M. (2013b). *Friday review: Dish of the day?* Accessed March 19, 2013, from http://www.totaltele.com

Lunden, I. (2013). *Sprint ups its offer for outstanding Clearwire shares to around $2.5B*. Accessed May 21, 2013, from http://www.techcrunch.com

Osawa, J. (2012). *SoftBank deal for Sprint turns on spectrum*. Accessed October 21, 2012, from http://www.online.wsj.com

Phone Factor. (2008). *A visual guide to 700 MHz*. Accessed October 21, 2008, from http://www.phonescoop.com

Ramachandran, S. (2012). *Dish frustrated in dream of breaking into wireless*. Accessed November 20, 2012, from http://www.totaltele.com

Sahota, D. (2012). *FCC approves AT&T's spectrum transfer to T-Mobile*. Accessed April 27, 2012, from http://www.telecoms.com

Standeford, D. (2013a). *FCC launches incentive auction process and ponders spectrum caps*. Accessed February 3, 2013, from http://www.policytracker.com

Standeford, D. (2013b). *Spectrum usage rights: Stalled in Europe but emerging in US*. Accessed February 21, 2013, from http://www.policytracker.com

Standeford, D. (2013c). *FCC explains incentive auction process but how interested are broadcasters?* Accessed February 21, 2013, from http://www.policytracker.com

Standeford, D. (2013d). *US reports show disagreement over reuse of government spectrum*. Accessed February 21, 2013, from http://www.policytracker.com

Stynes, T. (2012). *Clearwire, Leap Wireless reach LTE wholesale pact*. Accessed March 14, 2012, from http://www.totaltele.com

Tadena, N. (2011). *Moody's cuts ratings on Sprint, Clearwire*. Accessed October 17, 2011, from http://www.cellular-news.com

Telecom.paper. (2011a). *Sprint, Clearwire sign new wholesale agreement*. Accessed April 21, 2011, from http://www.telecompaper.com

Telecom.paper. (2011b). *Verizon buys mobile spectrum from cable cos for USD 3.6 bln*. Accessed December 5, 2011, from http://www.telecompaper.com

Telecom.paper. (2011c). *AT&T abandons bid for T-Mobile USA*. Accessed December 20, 2011, from http://www.telecompaper.com

Telecom.paper. (2012a). *FCC to block LightSquared network roll-out*. Accessed February 16, 2012, from http://www.telecompaper.com

Telecom.paper. (2012b). *Sprint backs out from MetroPCS at last hour*. Accessed March 7, 2012, from http://www.telecompaper.com

Telecom.paper. (2012c). *Verizon to market LTE for home broadband*. Accessed March 7, 2012, from http://www.telecompaper.com

Telecom.paper. (2012d). *FCC to vote on incentive auctions proposal*. Accessed September 12, 2012, from http://www.telecompaper.com

Telecom.paper. (2012e). *Dish launches dishNET satellite broadband service*. Accessed September 28, 2012, from http://www.telecompaper.com

Telecom.paper. (2013a). *Dish tops SoftBank bid for Sprint*. Accessed April 15, 2013, from http://www.telecompaper.com

Telecom.paper. (2013b). *Sprint increases offer for Clearwire*. Accessed June 20, 2013, from http://www.telecompaper.com

Telecom.paper. (2013c). *Dish withdraws Clearwire tender offer*. Accessed June 27, 2013, from http://www.telecompaper.com

Telecom.paper. (2013d). *AT&T to acquire Leap Wireless*. Accessed July 14, 2013, from http://www.telecompaper.com

TeleGeography. (2011a). *LightSquared and Sprint ink 15-year network sharing deal*. Accessed June 22, 2011, from http://www.telegeography.com

TeleGeography. (2011b). *Kentucky fried LTE? Appalachian Wireless confirmed as Verizon's twelfth rural LTE partner*. Accessed October 12, 2011, from http://www.telegeography.com

TeleGeography. (2011c). *LightSquared refutes report claiming 4G network interferes with GPS*. Accessed December 19, 2011, from http://www.telegeography.com

TeleGeography. (2012a). *Sprint confirms use of iDEN spectrum for LTE by 2014*. Accessed April 16, 2012, from http://www.telegeography.com

TeleGeography. (2012b). *Pioneer Cellular launches rural LTE network in Oklahoma*. Accessed May 5, 2012, from http://www.telegeography.com

TeleGeography. (2012c). *LightSquared prospects look dim as bankruptcy deadline looms*. Accessed May 14, 2012, from http://www.telegeography.com

TeleGeography. (2012d). *DISH does not expect LTE-Advanced until 2016*. Accessed May 23, 2012, from http://www.telegeography.com

TeleGeography. (2012e). *Copper Valley jumps on Verizon rural LTE bandwagon; Alaska set for LTE by end-2013*. Accessed June 20, 2012, from http://www.telegeography.com

TeleGeography. (2012f). *Sprint to fire LTE starting pistol in Atlanta, Dallas, Houston, Kansas City, San Antonio on 15 July*. Accessed June 28, 2012, from http://www.telegeography.com

TeleGeography. (2012g). *AT&T poised to acquire NextWave Wireless: Lines up series of 700MHz acquisitions*. Accessed August 3, 2012, from http://www.telegeography.com

TeleGeography. (2012h). *64 companies interested in Verizon's 700MHz spectrum*. Accessed August 21, 2012, from http://www.telegeography.com

TeleGeography. (2012i). *FCC votes to approve Verizon spectrum deals*. Accessed August 24, 2012, from http://www.telegeography.com

TeleGeography. (2012j). *Sprint: 'We are not (just) in Kansas anymore'; cellco confirms LTE expansion to four new markets.* Accessed August, 31 2012, from http://www.telegeography.com

TeleGeography. (2012k). *AT&T also targeting Comcast, Horizon WCS frequencies.* Accessed September 6, 2012, from http://www.telegeography.com

TeleGeography. (2012l). *Another one bites the dust! New rural WiMAX firm closes down.* Accessed September 7, 2012, from http://www.telegeography.com

TeleGeography. (2012m). *AT&T sets its sights on CenturyTel spectrum.* Accessed September 18, 2012, from http://www.telegeography.com

TeleGeography. (2012n). *Satellite company DISHes the dirt on 'self-serving' Sprint claims.* Accessed September 26, 2012, from http://www.telegeography.com

TeleGeography. (2012o). *CCA urges regulatory rethink over 'devastating' AT&T spectrum purchases.* Accessed September 28, 2012, from http://www.telegeography.com

TeleGeography. (2012p). *FCC paves the way for AT&T to utilise WCS frequencies for LTE.* Accessed October 19, 2012, from http://www.telegeography.com

TeleGeography. (2012q). *US LTE round up.* Accessed October 30, 2012, from http://www.telegeography.com

TeleGeography. (2012r). *Sprocket completes 'Phase o' of LTE deployment.* Accessed November 12, 2012, from http://www.telegeography.com

TeleGeography. (2012s). *FCC: DISH must roll out 40% of its LTE network in four years.* Accessed December 19, 2012, from http://www.telegeography.com

TeleGeography. (2013a). *AT&T, Verizon, T-Mobile to explore spectrum sharing with government.* Accessed February 1, 2013, from http://www.telegeography.com

TeleGeography. (2013b). *T-Mobile wades into spectrum debate, supports 600MHz interoperability.* Accessed February 5, 2013, from http://www.telegeography.com

TeleGeography. (2013c). *Better late than never? T-Mobile will finally launch LTE this month.* Accessed March 19, 2013, from http://www.telegeography.com

TeleGeography. (2013d). *Sprint blames vendors for slow network vision progress.* Accessed May 8, 2013, from http://www.telegeography.com

TeleGeography. (2013e). *DISH lodges USD2bn 'stalking horse' bid for LightSquared.* Accessed May 21, 2013, from http://www.telegeography.com

TeleGeography. (2013f). *DISH asks the FCC to waive 700MHz E block rollout requirements; withdraws Clearwire bid as wireless options shrink.* Accessed June 27, 2013, from http://www.telegeography.com

TeleGeography. (2013g). *FCC sets wheels in motion for 1900MHz H block auction.* Accessed June 28, 2013, from http://www.telegeography.com

TeleGeography. (2013h). *Verizon LTE network reaches 500 markets; PCS spectrum refarming slated for 2015.* Accessed June 28, 2013, from http://www.telegeography.com

Wakabayashi, D. (2012). *SoftBank in advanced talks to acquire Sprint.* Accessed October 11, 2012, from http://www.totaltele.com

Wood, N. (2013). *T-Mobile, MetroPCS merger closes.* Accessed March 19, 2013, from http://www.totaltele.com

Yu, R. (2013). *Dish raises bid to buy Clearwire to top Sprint's offer.* Accessed May 30, 2013, from http://www.usatoday.com

India, Russia and the UK

4

4.1 India

The situation in India where spectrum allocation is concerned can best be described as an ongoing mess, the full extent of which is too complicated to detail here. Suffice it to say that India, as noted below, has managed within a period of a mere 2 years to issue 2G, 3G and (effectively) 4G licences within the context of a complex regionally-based licensing system with the result that—subject in part to the resolution of ongoing disputes—it remains unclear which operators will be providing which technologies in which parts of India. In particular, as was admittedly also true of certain other countries in Asia—see, for example, Bangladesh and Pakistan in Chap. 5—it was still in the throes of discussing how to allocate 3G spectrum at a time when Europe and the USA were considering the issue of 4G licences, and as a result the government decided to hold back-to-back auctions for 3G and BWA licences (Government of India 2010).

The 3G spectrum, comprising 5 MHz paired blocks in the 2.1 GHz band, was offered independently in all 22 'circles' (main cities/regions) in April 2010. There were nine qualified bidders: Aircel, Bharti Airtel, Etisalat, Idea Cellular, Reliance (RCom), S-Tel, Tata Teleservices, Vodafone Essar and Videocon Telecoms—Uninor (subsequently Telenor) and Loop were not interested. The auction ended after 34 days and 183 rounds and INR 677 billion ($14.63 billion) was raised including the matching prices paid by state-owned incumbents BSNL and MTNL which between them covered the whole of India. No bids were made for a licence in every circle—no surprise given the huge cost of so doing. Aircel won in 13 circles at a cost of $1.40 billion; Bharti Airtel won in 13 circles at a cost of $2.66 billion; Reliance won in 13 circles at a cost of $1.86 billion; Idea won in 11 circles at a cost of $1.25 billion; Vodafone Essar won in 9 circles at a cost of $1.74 billion; and Tata Teleservices won in 9 circles at a cost of $878 million. S-Tel won cheaply in 3 circles but Etisalat and Videocon were unsuccessful. On 1 September, the 3G spectrum was allocated on time.

P. Curwen and J. Whalley, *Fourth Generation Mobile Communication*,
Management for Professionals, DOI 10.1007/978-3-319-02210-9_4,
© Springer International Publishing Switzerland 2013

On 23 May 2010, the BWA (2.3 GHz) auction commenced in all circles but with the potential for two nationwide licences to be acquired. The spectrum was unpaired and divided into 20 MHz bands. There were 11 qualified bidders: Aircel, Augere, Bharti Airtel, Idea Cellular, Infotel Broadband Services, Qualcomm, Reliance WiMax (RCom), Spice ISP, Tata Communications Internet services, Tikona Digital Networks and Vodafone. The auction ended after 16 days and 117 rounds. RCom had exited early and the spectrum was assigned to Infotel Broadband Services which won a nationwide licence (intending to use WiMAX), Bharti Airtel which won spectrum in four circles, Aircel which won in eight circles, Tikona Digital Networks which won in five circles and Augere which won in one circle and pencilled in a launch for April 2012 (the latter both favouring WiMAX). Those not already in possession of a unified licence were initially forbidden to provide voice services, but this ban was lifted in February 2013 (Telecom.paper 2013a).

For its part, Qualcomm (the majority shareholder in a consortium including Global Holding Corp. and Tulip Telecom) won in four circles including Delhi and Mumbai at a cost of $1.05 billion and announced its intention to use TD-LTE even though the spectrum being used in China for TD-LTE was different and hence there were no available devices. In October 2011, it offered the licences for sale, either all four together for no less than what it had paid or in two packages of two licences based on Delhi and Mumbai, with a proviso that LTE had to be used (Cellular-news 2010). However, it subsequently announced that it would roll out a network in late 2011 before looking for a purchaser—initially for the 26 % stake held by Global Holding and Tulip Telecom. In practice, the licences were handed over only in May 2012 after a lengthy dispute and with the original termination date still in force (Telecom.paper 2012e). Qualcomm promptly agreed to sell the 26 % owned by Tulip/Global to Bharti Airtel which contracted to raise this stake to 49 % at a total cost of $165 million. Once the network is in commercial operation, Bharti will acquire the rest of the shares (Telecom.paper 2012f)—it bought an additional 2 % in July 2013.

$5.48 billion was raised in total, but adding in the amounts paid by state-owned BSNL and MTNL—both initially committed to WiMAX but now leaning towards LTE—which had been guaranteed spectrum in advance of the auction at equivalent prices to those paid by the private operators, raised this to $8.22 billion (Gabriel 2010). Reliance Industries (to be distinguished from the previously divested operator RCom) immediately agreed to buy 95 % of Infotel for $1 billion, thereby creating Reliance Jio Infocomm. It initially opted to develop LTE, but with no prospects of coming to market for a year or so, decided to trial WiMAX in order to get to market much earlier (Kinetz 2010). The latest estimate for the launch of LTE is end-2013 by which point it may have sold a 25 % stake to AT&T.

It was assumed that the other winners would also opt for WiMAX—at least initially—because of the need to launch services quickly in a competitive market. However, Bharti Airtel's membership of the GTI may have dealt a death blow to the prospects for WiMAX in India—it contracted with ZTE to roll out TD-LTE in October 2011—which were further damaged by successful trials of TD-LTE carried out by Aircel in August 2011. MTNL's failed attempt to attract a

WiMAX licensee led it to re-offer the franchise on a technology-neutral basis in August 2011 (TeleGeography 2011a). BSNL had intended to roll out TD-LTE in cities even though it was obliged to stick with WiMAX in rural areas. However, in October 2011, it announced that the licence was too expensive and that it had made negligible progress in utilising its spectrum despite its head start over the private operators, and hence that it wished to return its licence (Cellular-news 2011b). In mid-December, it duly handed back spectrum in 9 of its 20 circles.

The accepted wisdom that, although the market was highly competitive with some consolidation on the cards, the recent recipients of 3G licences would want to move on to LTE was thrown into disarray in February 2012 when the Supreme Court ruled that the 122 2G licences awarded in 2008 had been awarded illegally and would be revoked in June (subsequently extended to January 2013). This would result in the need to re-issue the licences in January 2013 after successive auctions for 1,800 MHz GSM and 800 MHz cdma2000 licences, the disappearance of some of the 2008 new entrants (with connections to large international operators) (TeleGeography 2012r) and a radical re-appraisal of the value of LTE to those operators left standing. The decision by the regulator that 3G roaming would also be disallowed was also bound to be a deterrent to LTE take-up (Lennighan 2012a).

Those operators not caught up in the problem were likely to take such advantage as they could to gain a first-mover premium, and Bharti Airtel, for example, was quick to sign a deal with NSN to roll out a TD-LTE network in Maharashtra. However, it subsequently announced that it hoped to achieve its first launch in Kolkata, where the vendor was ZTE, before the end of March, with the launch taking place in early April (Telecom.paper 2012b). Aircel was expected to follow suit, but announced in October that it would not be launching until 2013 due to market and regulatory uncertainty (TeleGeography 2012s).

In early March, the so-called Empowered Group of Ministers announced that the 700 MHz band would be allocated for 4G service provision. They added that they had asked the Ministry of Defence to vacate 230 MHz of the 300 MHz of spectrum that it held, but that the Ministry was holding out to retain 150 MHz (Telecom.paper 2012a). This was followed by the regulator's consultation paper concerning the re-auctioning of the revoked spectrum and associated matters (Telecom Regulatory Authority of India 2012).

In the run-up to the auction reassigning 2G spectrum, various agencies—and there are more of these (political, bureaucratic as well as legal) seeking to interfere in telecommunications than in other countries—gave out a whole series of mixed messages about such matters as re-farming and spectrum swaps [see, for example, TeleGeography (2012p)]. However, it does appear to be the case that spectrum-sharing will be permitted in specified circumstances (Telecom.paper 2012i).

What bearing all this will have on the provision of 4G remains to be seen. The 2G auction certainly suggests that operators intend to proceed with caution. In response to a minimum price per 1.25 MHz block of 1,800 MHz band spectrum set many times higher than in 2008, the interest in the auction in this band in November could best be described as muted—it was already known that neither Bharti Airtel, nor any other pre-qualified operator bar Idea Cellular, was interested in bidding for

2G spectrum in every circle (Telecom.paper 2012j). In the event, only five operators participated. Telenor (Telewings) and Videocon, both of which had lost all of their licences, together with Idea Cellular whose seven licences had been revoked, were effectively forced to bid in order to stay in business—although other 2G new entrants had chosen either, like Batelco, to cease to operate altogether or, like Sistema Shyam, to sue for the return of its licences. The other bidders, Vodafone and Bharti Airtel, simply needed additional bandwidth to overcome QoS issues. Not surprisingly, bidding was initially lethargic, but at least some offers were tabled in comparison to the situation in respect of the CDMA-based licences where no bids at all were forthcoming—the minimum price was even higher than for GSM bandwidth—and the auction had to be cancelled (Krishna 2012a).

The GSM auction quickly petered out, lasting only 2 days and raising only $1.71 billion with all bidders acquiring some spectrum (Krishna 2012b): Idea won five blocks in one circle, four blocks in six circles and one block in one circle; Vodafone—restricted to either one or two blocks per circle—won two blocks in nine circles and one block in five circles (as it was unaffected by cancelled licences it managed to bid as little as $217 million); Telenor won four blocks in six circles at a cost of $720 million (although it had previously held licences in nine circles); Videocon won five blocks in six circles at a cost of $404 million (subsequently announcing that it would roll out its network during 2013H2); and Bharti Airtel won one block in one circle. No bids were entered for either Delhi or Mumbai (Telecom. paper 2012l) and, overall, licences in only 35 circles of the 122 revoked were sold, with 20 circles generating 85 % of the revenue from operators that had lost licences. In the aftermath, it was claimed that all unsold GSM and CDMA licences would be re-auctioned before March 2013, but this appeared to run foul of a legal restriction on any attempt during the current fiscal year to adopt the sensible strategy of lowering the reserve prices (TeleGeography 2012t).

In December, the Empowered Group of Ministers announced that a new auction of unsold GSM—but not of unsold CDMA—spectrum would take place in March 2013 with minimum prices set 30 % below those set at the initial auction (Cellular-news 2012e). Subsequently, it was determined that re-auctioned CDMA spectrum would be sold at half the previous minimum price and 900 MHz spectrum at twice the minimum price for 1,800 MHz spectrum (TeleGeography 2013c). Vodafone not merely complained loudly that the GSM minima were still set at too high a level but also noted that it had yet to receive its new licences which should accordingly be dated from when they were handed over as against when they were awarded—they were eventually handed over in June 2013. It also took legal action to prevent the re-auctioning of its 900 MHz licences upon expiry in Delhi, Mumbai and Kolkata on the grounds that it had the right to apply for a 10-year extension, but was unsuccessful. Its response was to offer $680 million for the right to retain the licences.

As 2013 dawned, many matters remained unresolved. For example, any GSM operator holding more than 4.4 MHz of bandwidth would be obliged to pay a surcharge as from October 2012, whereas if it held more than 6.25 MHz it would be obliged to choose either to surrender any excess spectrum or to pay a retroactive penalty fee covering the period July 2008 to October 2012. Furthermore, any (presumably foreign) company that tried to buy an incumbent despite having declined to participate in the 2G auction would have to pay a surcharge. In addition, licences in the 900 MHz band would come up for renewal in 2014, and this was expected to lead to significant re-farming of the band (Telecom.paper 2012k). The effect this could potentially have on an operator's 2G spectrum holdings was clearly one of the factors causing operators to proceed with caution during the 1,800 MHz auction.

In February 2013, Sistema Shyam (MTS India) shut down its operations in 10 circles while at the same time confirming its intention to bid in the forthcoming auction in March. As it turned out, it was the only operator to lodge an intention to bid but only for spectrum in the 800 MHz band—the 900 MHz and 1,800 MHz spectrum was simply ignored (Telecom.paper 2013b). It duly paid $668 million for eight licences, each providing 4.75 MHz of bandwidth, ending up after other adjustments with nine of its original 12 licences. Both Sistema and Telenor were permitted to offset the licence fees paid in 2008 against the fees paid in 2012, which still meant a net payment needed to be made whereas the offset granted to Videocon and Idea Cellular meant that were allowed to offset their entire 2012 fees (TeleGeography 2013i).

In March, Videocon stated that it would be launching LTE in six circles by the year-end. However, the outlook for operators remained cloudy. For example, Bharti Airtel was ordered by the DoT to cease offering a 3G service (via roaming) in the seven circles where it did not hold a 3G licence—it promptly appealed to the Delhi High Court which ruled that it could continue roaming but must no longer sign up new subscribers outside its licensed circles—a ruling subsequently extended to include Vodafone and Idea Cellular. Secondly, CDMA operators were ordered to pay one-off fees for any spectrum held beyond that allocated to them when they originally launched—that is, anything in excess of 2.5 MHz in the 800 MHz band— or forfeit the spectrum. This ruling affected Sistema, RCom, Tata, BSNL and MTNL but did not apply to Sistema's newly acquired licences. In response, Tata agreed to hand back its excess spectrum in all circles bar Delhi and Mumbai. Thirdly, the Comptroller and Auditor General accused operators of acting as a cartel in the recent auctions, claiming that the major operators had agreed among themselves not to participate, and also of overstating their subscriber numbers such as to give the false impression that they were short of spectrum.

The prospects for LTE were somewhat dampened in March when it was revealed that so far only 2 % of the population had signed up for 3G. Sistema subsequently announced that it wanted to get its finances onto a more secure footing before launching LTE, while, for its part, the Trai recommended that India should adopt the APT band plan for the 700 MHz band utilising 45 MHz paired of spectrum. On a slightly more positive note in April, Reliance Jio Infocomm was awarded 10,000 numbers divided between three cities in order to test LTE. Also positive, at least in principle, was the announcement in July that, subject to Cabinet approval, the limit on Foreign Direct Investment in the sector would be raised from 74 % to 100 %, of which 49 % could be bought directly and more added later subject to approval by the Foreign Investment Promotion Board. However, this seems unlikely to herald a stampede of new investors given the treatment meted out to those who entered at an earlier date. Indeed, the need for consolidation is on everyone's minds—even the Trai came out in favour in July.

4.2 Russia

What follows is the authors' best effort to disentangle a very confusing story. It treats MTS, VimpelCom and MegaFon as the three national incumbents, but in this case it is the unusual number of other actual or potential operators that makes it difficult to create a simple historical narrative. The most sensible point to begin seems to be when the Ministry of Defence acquired a 25.1 % stake (via Voyentelekom) in a new (potentially national) LTE operator registered as Osnova Telecom. It was awarded some 2.3 GHz and 2.4 GHz spectrum in order to set up a trial network, but the MoD also proposed that Osnova would collect together the licences in the 2.3 GHz and 2.4 GHz bands won primarily by Rostelecom (essentially a fixed-wire operator) during four regional tenders in February/March 2010.

To forestall this, the three national operators, MTS, MegaFon and VimpelCom, together with Rostelecom, decided to attempt a short-cut similar to the model espoused by LightSquared in the USA (see Chap. 3). In this case, a WiMAX operator owned by Scartel, but operating as Yota Networks, would switch from WiMAX to LTE in 2011 and then provide a wholesale service to the four operators. Scartel's licences would enable the provision of both FDD and TDD spectrum on a near-national basis by 2014, at which point each operator would be expected to take a 20 % stake in Scartel (Middleton 2011). However, three large regional operators were excluded from joining the consortium and their complaints meant that the matter ended up in the hands of the anti-trust agency. This was not expected to matter in practice given that Scartel placed a value on itself of $1 billion in August 2011, which the operators did not consider to be acceptable. Nevertheless, Tele2 was determined not to be excluded from LTE provision and it accordingly launched in September an independent investigation into the possibility of re-farming the 1,800 MHz spectrum band. Tele2's situation is unusual in that it has admitted that its prospects of being awarded a 3G licence are minimal, and hence it must skip from EDGE+ to LTE if it is to compete with its three larger rivals (Grundberg 2012). In March 2012, it announced that it was about to test LTE using re-farmed 1,800 MHz spectrum in Omsk. Subsequently, there were rumours of a potential merger with Rostelecom but the reality turned out to be a sale to the VTB Group, completed in a mere 7 days in early April 2013 in the face of a much higher counter-offer tabled by MTS and VimpelCom.

Also in September, much of the above was thrown into question when MTS and VimpelCom announced that they intended to roll out a jointly-funded network while Tele2 sought permission in October to operate a trial in two areas which it struggled to secure. In practice, matters are rarely straightforward in Russia but the situation in Moscow at the time was particularly complex, involving possible exchanges of WiMAX and LTE spectrum, various subsidiaries of the mobile operators and the military—see, for example, TeleGeography (2011b). Provided all went well, and spectrum suitable for LTE was aggregated in the hands of the regulator, it would be auctioned off, but recent political developments indicated further delays.

In November, the Ministry of Defence finally authorised the building of a network using spectrum in the 2.3 GHz and 2.4 GHz bands, but granted Rostelecom rather than Osnova the right to be the network builder (TeleGeography 2011d). Where this network had no coverage, both Rostelecom and MegaFon would act as MVNOs on the Scartel LTE network which launched on 10 April 2012 (Telecom. paper 2012c). This ostensibly gave Rostelecom and MegaFon a head start over VimpelCom and MTS (TeleGeography 2011c), although the fact that Rostelecom and MTS signed a fixed/mobile network-sharing agreement in February 2012 suggested that matters were rather more complex in practice. Indeed, further complexity was added in late April 2012 when AF Telecom—the holding company of Alisher Usmanov which owns 50 % + 1 share in MegaFon—and Scartel announced the setting up of a joint venture, subject to regulatory permission, to be owned 80 % by AF Telecom and 20 % by Scartel (Telecom.paper 2012d) which is itself 74.9 % owned by a private investment fund, Telconet Capital, and 25.1 % by Russian Technologies State Corporation. The Federal Antimonopoly Service approved the joint venture in June 2012 with the proviso that a wholesale service should be provided.

Irrespective of how that turns out, it was evident that MegaFon was pursuing a two-pronged strategy: firstly to launch as a MVNO on the Scartel network— initially achieved in late-April 2012 in Novosibirsk followed by Moscow in mid-May (TeleGeography 2012i); secondly, to launch in its own right using the 300 base stations built by mid-May. However, the latter project, which will probably use TD-LTE, was hamstrung by the need to give up the spectrum in the 2.57–2.62 GHz band in the hands of its subsidiary Synterra so that it could be auctioned off at some point (TeleGeography 2012e).

Meanwhile, in December 2011, Osnova announced that it now intended to secure spectrum in the 790–862 MHz band for its proposed LTE network which it would use in cities. Osnova and Scartel may not end up as the sole wholesalers since a further company, a WiMAX operator owned by Prestige Internet called Enforta with spectrum holdings in the 2.5 GHz and 3.5 GHz bands, announced in March 2012 that it was about to begin testing LTE in the (somewhat unsuitable) 3.5 GHz band (TeleGeography 2012c). However, it was reported in July 2013 that Enforta had agreed to relinquish its holdings in the 2,530–2,570 MHz and 2,650–2,690 MHz bands in 102 cities—what it would get in return was not specified—so its future role is unclear.

However, the real mystery surrounds a company identified as Antares—see also Vietnam in Chap. 5—which is alleged to intend rolling out a network in Moscow using the 1,900–1,920 MHz band. It is alleged that Antares, together with sister companies Arktur and Integral, were awarded 5-year LTE licences as long ago as

April 2008. The regional licences were distributed as follows: Antares—17 regions in central Russia including Moscow and seven regions in Siberia/Urals; Arktur—14 regions in the Volga, the Urals and Bashkortostan and 10 in north-west Russia/St Petersburg; Integral—13 regions in North Caucasus, nine in the Far East Federal District and 12 in Siberia (TeleGeography 2012d). Antares was given permission to use the spectrum for LTE in May 2012. In February 2013, the Moscow Arbitration Court postponed making a decision as to whether Antares should be allowed to hold on to the spectrum beyond its termination date of April 2013, but subsequently found the State Radio Frequency Commission to be 'guilty of inaction' and demanded a quick resolution of the licence issue—which is unlikely to transpire as the SRFC can appeal the decision (TeleGeography 2013g).

The confusion over licences deepened when MTS announced in February 2012 that it had been awarded a LTE licence covering Moscow and the surrounding region, valid until end-2016 with a mandatory launch date of end-2013. The spectrum in the 2,595–2,620 MHz band would be used for TD-LTE (TeleGeography 2012a). However, it appeared that this spectrum had been taken back from former MTS subsidiary Comstar by the regulator, Roskomnadzor, so that it could issue licences on a nationwide basis—or possibly ten regional licences (TeleGeography 2012g). As a result, compensation would have to be paid to MTS and also to Kosmos-TV, both of which are subsidiaries of Sistema. The need to determine the relevant amount seemed set to delay the auction in this band until September 2012. Meanwhile, MTS declared its intention to launch in Moscow on 1 September, so who was going to end up with licences, and who (if anyone) would end up paying compensation and to whom, was something of a mystery (TeleGeography 2012k). In late August, it was announced that the spectrum would be distributed among MegaFon, VimpelCom, Rostelecom and Scartel, with each paying $8.5 million to Kosmos-TV by way of compensation (TeleGeography 2012m). For its part, MTS launched in Moscow and 40 towns in nearby provinces in September 2012, adding that it expected to have coverage across 96 % of Moscow by the year-end (Cellular-news 2012c).

In June 2012, Scartel had admitted that it had not had sufficient time to optimise the switch from WiMAX to LTE, and that the switch-over in Moscow had been done on a once-and-for-all basis without any previous testing. Somehow, despite difficulties in arranging for WiMAX customers to obtain LTE dongles and fairly poor transmission speeds, it had gone smoothly, but now that the operator knew that this strategy could succeed it was embarking on an ambitious expansion programme (TeleGeography 2012j) and testing LTE-Advanced with Huawei. Scartel was estimated to have 600,000 LTE subscribers at end-August 2012. As of early September, the only place in Russia able to receive LTE signals from three operators—Scartel, MTS and MegaFon—was Kazan, the capital of the Republic of Tatarstan.

However, a crucial factor was that the Scartel network was set up to handle only four MVNOs, which were assumed to be MTS, MegaFon, VimpelCom and Rostelecom. But with Scartel launching in its own right as Yota, that only left room for three others. With VimpelCom receiving permission to launch over the network in September—it signed up formally as a MVNO in December (Telecom. paper 2012m)—this indicated that Rostelecom would be the one to lose out. As a result, Rostelecom cancelled its plans to use the Scartel network on an indefinite basis—launching a case in the Moscow Commercial Court (TeleGeography 2012n).

As for the 791–862 MHz band, Osnova declared in February 2012 that it had been awarded a licence and expected a full launch to take place in 20 cities in October or November 2013. It was claimed that the network would be built by Infra-Engineering, a company linked to Rostelecom (TeleGeography 2012b) but recent reports indicate that Huawei and Alcatel-Lucent are the vendors although it is unclear which spectrum band is involved. Notwithstanding this, the regulator set out to allocate (without a fee but with a minimum investment obligation of $458 million annually) four 7.5 MHz paired blocks with nationwide coverage in July 2012. Only one block per operator was permitted with preference given to any operator willing to host MVNOs on its network. Successful operators would be given 7 years to achieve coverage of all towns with 50,000 inhabitants (TeleGeography 2012f). Each licensee would also receive—when available— 7.5 MHz paired in the 720–791 MHz band and 10 MHz paired in the 2.6 GHz band. MTS, VimpelCom and MegaFon duly applied to bid, as did Summa Telecom which had previously had to abandon an attempt to roll out a nationwide WiMAX network (TeleGeography 2012h). The other bidders were Rostelecom, TransTeleCom (TTK) and two Tele2 subsidiaries (Omsk and Voronezh). MegaFon, MTS, Rostelecom and VimpelCom duly won the licences as expected and now have to compensate the previous licence holders. Rostelecom, which had been awarded 791–798.5 MHz paired with 832–839.5 MHz, announced in August 2012 that it would be launching in the North Caucasus federal district in 2013 (TeleGeography 2012l), while VimpelCom, which had been awarded 813.5–821 MHz paired with 854.5–862 MHz, was allegedly authorised to launch in Moscow in January 2013 (TeleGeography 2013b). As to when some of the spectrum won will be vacated by the military and broadcasters, that is another matter entirely. It was claimed that the military still held all the 791–862 MHz band spectrum in July 2013 (but see below).

However, the situation is by no means as straightforward as it seems because immediately after the results were announced, a court in Moscow ruled that the State Radio Frequency Commission (SRFC) had illegally rejected Summa Telecom's application for 70 MHz of spectrum in the 2.6 GHz band, a good part of which had previously been allocated to Scartel with an additional element sold in the auctions (Telecom.paper 2012h). The Ninth Arbitration Court of Appeals promptly over-ruled this decision and the matter is rapidly turning into yet another Russian legal drama (TeleGeography 2013a). In the latest move, the Federal Arbitration Court upheld the decision of the lower court in February 2013 (TeleGeography 2013h).

In February 2013, the new Defence Minister offered to return Osnova's LTE frequencies because it was 'inexpedient' to roll out a LTE network (TeleGeography 2013d). However, Osnova's largest shareholder, Aikominvest, immediately stepped in to deny that the Defence Ministry had the right to do so (TeleGeography 2013e). For its part, VimpelCom stated that it would be concentrating upon its 3G roll-out for the time being, but nevertheless intended to launch in Moscow and five other cities during 2013.

MegaFon announced that it had acquired 130,000 LTE subscribers during 2012, and hoped to raise this total to one million during 2013. Elsewhere, controversy arose concerning whether equipment was being manufactured by companies majority-owned by Russian nationals as required by law. LTE base stations developed by the Wireless Technology Centre, owned by Micran (25 %) and NSN (75 %), for regional operator Vainakh based in the Chechen Republic were deemed by the regulator to be unacceptable for a commercial launch, and as a result the launch by Vainakh in January 2013 was deemed to be only a trial pending resolution of the ownership issue (TeleGeography 2013f). Further controversy then arose when an appeal by a regional MMDS operator, Eros, against the award of LTE frequencies to Scartel/Yota, was successfully pursued in an appeals court on the grounds that the Scartel service was interfering with its transmissions (Telecom. paper 2013c).

Scartel also found itself under investigation by the anti-trust authorities as a result of a complaint by Rostelecom to the effect that Scartel was discriminating against Rostelecom by denying it access on wholesale terms (Telecom.paper 2013d).

Under the terms of the 2012 auction, the four successful bidders were obliged to deliver an initial launch by 1 June 2013, and as a result LTE has become increasingly available. VimpelCom launched in parts of Moscow at the end of May using the 800 MHz and 2.6 GHz bands (TeleGeography 2013k) with MTS and MegaFon following suit in Moscow and Yekaterinburg respectively, whereas Rostelecom launched in Sochi on 1 June (TeleGeography 2013l) and selected a further seven regions for the next stage of its launch. These launches are to be distinguished from where operators had previously launched as MVNOs.[1]

The stability of the Russian market came into question yet again when it was alleged in June that MegaFon intended to take over Scartel. Given that the companies are both part of the Garsdale holding company with a common controlling shareholder in Alisher Usamov this is not an entirely surprising development, and the deal has been approved by the anti-trust authorities. It was further alleged that the government would sell its 55.6 % stake in Rostelecom to the 'big three', in effect snubbing foreign investors. At the same time, VTB Group announced its intention to dispose of a controlling stake in Tele2 with Rostelecom cited as the likely buyer. So, in effect, Rostelecom would obtain control of Tele2 before being split up between VimpelCom, MTS and MegaFon, the first two of which had offered to buy Tele2 at a much higher price than VTB Group in the first place. All rather curious, but things will doubtless turn out rather differently in

---

[1] Ongoing commentary can be found in the Broadband and Satellite Russia Newsletter published bi-monthly by the Russian Satellite Communications Company (http://www.rscu.ru).

practice, especially as VTB Group have stated that it is not their intention to sell to Rostelecom.

Despite all the confusing developments in other areas, the one positive aspect in July 2013 appears to be the launch of the LTE networks by the incumbents acting on their own behalf rather than as MVNOs. However, even that apparently may no longer be the case as the Ministry of Communications is alleged to have tabled a proposal that will strip the licences from the four operators that met the 1 June 2013 deadline for an initial launch on the grounds that they are proceeding far too slowly, building over-lapping networks, using only 5 MHz paired of the 7.5 MHz paired awarded and concentrating on high-income customers in big cities. Claiming that a single common network requires the construction of 30,000 base stations, the Ministry claims that the only way to achieve that target within a reasonable time period is to re-allocate to itself spectrum in the 720–750 MHz and 761–862 MHz bands together with any spectrum that subsequently becomes available in the 390–470 MHz and 694–876 MHz bands. It is alleged that it proposed to deal with this by transferring the spectrum to a state-owned company which would regulate the prices that could be charged by private operators acting as MVNOs (Lennighan 2013). However, the Ministry has been quick to deny all such allegations although it has reiterated its desire for more rapid progress.

4.3 UK

Arqiva has conducted trials in Wales using digital dividend spectrum as has Clear Mobitel in Cornwall. However, the first equivalent move by an operator, also in Cornwall, involved Everything Everywhere (in conjunction with BT Wholesale) using two 10 MHz sections of the digital dividend band, commencing in October 2011 and running until June 2012. In November, Telefónica (O$_2$) announced that it had been awarded a Test and Development licence for the 2.6 GHz band and would be conducting a trial in central London over the coming half-year. This would supplement the previous trials in Slough which commenced in 2009. Hutchison 3G UK (H3G) proposed to launch a trial in the Thames Valley in March 2012.

In June 2011, the owner of a nationwide licence in the 3.5 GHz band, UK Broadband, which had been authorised by the regulator (Ofcom) to provide mobile as well as fixed-wireless services, announced that it was intent upon rolling out a LTE network as a wholesale provider, commencing in the main urban centres in 2012. In practice, this spectrum is probably uneconomic for LTE, especially if competitors are using the 800 MHz band.

Despite the above indications of progress, official policy was in something of a mess. It was intended that 250 MHz of digital dividend and 2.6 GHz band spectrum would be auctioned simultaneously, hopefully in 2012Q1—subsequently post-poned until 2012Q4. Commercial roll-outs were accordingly likely to be delayed until 2013 or even 2014 due to a combination of legal challenges, interference issues and other problems. The regulator clarified the position in March 2011 (Ofcom 2011), emphasising that the main objective was to secure the continuation of the same operator structure as already existed, which meant that any outcomes not achieving that aim would be disregarded. There would be spectrum caps and floors both in the sub-1 GHz bands and overall (House of Commons Culture and Media and Sport Committee 2011).

In September 2011, the regulator was forced to push back the auction schedule by at least one quarter as a result of, firstly, allegations that a spectrum cap on digital dividend spectrum could be viewed as illegal state aid for H3G, which would be able to acquire a relatively large chunk of the spectrum on offer and, secondly, the need to deal with the relocation of Digital Terrestrial TV—known as Freeview—which was accessed by a large proportion of households (Cellular-news 2011a). In October, it announced that further consultation with potential bidders would be needed in the light of responses to its previous publications, with the outcome not being revealed until 2012Q3.

In November, the Ministry of Defence announced that it was open to short-term spectrum sharing in addition to the sections of 2,310–2,400 MHz and 3,410–3,600 MHz bands that it intended to release in 2013/2014 and 2015/2016 respectively. The MoD is likely to apply for a Crown Recognised Spectrum Access which it will then trade to a civilian partner under the Trading Regulations. However, the situation is somewhat ambiguous because if the proposal involves the MoD giving third-parties use of its spectrum for a specified period of time for a fixed fee then this is tantamount to a lease.

In January 2012, Ofcom announced that 4G would have to be made available to 98 % of the population—requiring one operator using digital dividend spectrum to cover areas not covered by 2G—and that it was removing the spectrum caps/floors that appeared to favour Everything Everywhere and H3G (Cellular-news 2012a).

This complex situation was compounded by the controversy over the quarter-share of their joint 1,800 MHz spectrum that T-Mobile and Orange were obliged to return to the regulator via a condition imposed by the European Commission when authorising their merger to form Everything Everywhere in March 2010. Everything Everywhere wanted Ofcom to restrict the ability of Vodafone and Telefónica to acquire this spectrum as they were the only operators with existing 900 MHz licences. The latter argued, however, that the 1,800 MHz spectrum could be used for 4G pending the release of alternative spectrum bands, and hence that any restriction would be unfair (Gabriel 2012a). There was a further issue in that if Everything Everywhere were to sell the spectrum to a new entrant, Ofcom's previous move to ensure favourable treatment in the digital dividend auction for the smallest remaining operator, H3G, would no longer apply, but Ofcom was anyway forced to rescind its specific protection for H3G in January 2012 when faced with further legal action threatened by Vodafone and Telefónica, although the caps and floors would remain in place (Gabriel 2012b).

In late February, Everything Everywhere announced that it would re-farm its 1,800 MHz spectrum used for 3G in order to launch LTE, and that a trial would begin in Bristol in April—a small trial in Threlkeld in Cumbria was also begun in June. It went on to announce in mid-March that Ofcom had provisionally given its approval for a launch using this spectrum before the end of 2012, with a final decision due in April. This provoked an outraged response from Vodafone, which argued that the largest player in the market had effectively been granted a head start over its rivals and would now have an incentive to bog down the 4G licensing process in litigation in order to maintain its lead (Cauchi 2012).

In a curious blog, H3G also announced the launch of 4G, but then revealed that this was actually DC-HSPA+ which it confusingly called HSPA+42 (meaning 42 Mbps downstream) (Smith 2012). It subsequently transpired that Everything Everywhere was negotiating with both H3G, with which it shared infrastructure, and Virgin Mobile, a large MVNO using its network which it also intends to do so in respect of LTE (Gabriel 2012d), to give the impression that it was not seeking simply to dominate 4G by itself. As a result of negotiations with the European Commission at the time the merger forming Everything Everywhere was sanctioned, both H3G and Virgin Mobile had the right to launch LTE 6 months after Everything Everywhere. But this posed a dilemma for Everything Everywhere because it needed to agree a wholesale contract with H3G before H3G could launch, and knowing that it was being wooed as a potential ally potentially put H3G in the driving seat. However, H3G was the most active purveyor of low-cost data packages, and there was a risk that it would try to undercut Everything Everywhere on the back of a favourable wholesale agreement.

Meanwhile, the debate over the need for special treatment for H3G in order to preserve a fourth operator in the market rumbled on, with H3G—having secured extra capacity via its upgrade to HSPA+42—threatening to sue Ofcom if it did not receive special treatment in respect of sub-1 GHz spectrum (Cellular-news 2012b) and Vodafone claiming that H3G was no longer a small new entrant requiring protection. In early May 2012, Hutchison Whampoa's Managing Director allegedly told the government that it was prepared to cut its future investment or even, as a last resort, pull out of the UK altogether if the playing field was not tilted in its favour during the 4G licensing process.

In August 2012, Ofcom eventually authorised Everything Everywhere to proceed with a LTE launch using 1,800 MHz spectrum at any time from September onwards, arguing that any further delays would be detrimental to the interests of consumers (BBC 2012), whereupon Vodafone issued a lengthy rebuttal of the arguments underpinning the decision. It was noted that handsets capable of utilising LTE in the 1,800 MHz band had yet to appear on the market. As part of the agreement with Ofcom, as noted above, Everything Everywhere then announced that it would transfer 15 MHz paired of the spectrum, with its attached right to be used for LTE, to H3G. However, it was given a year to clear the spectrum before handing it over—specifically, 10 MHz paired is to be handed over on 1 October 2013 and 5 MHz paired on 1 October 2015—so H3G will not be launching until the latter part of 2013. To that end, it contracted Samsung to roll out its network, the first time that the vendor had been involved in a full commercial roll-out in Europe.

Everything Everywhere was rebranded as 'EE' (Lennighan 2012b) and announced in September that it would provide coverage of 16 cities, accounting for one-third of the population, before the year-end, subsequently specifying the launch date for the first 11 cities as 30 October. It claimed that it would be investing

£1.5 billion—a high-end amount perhaps boosted by the imminent launch of the iPhone 5 able to operate on the 1,800 MHz band—which raised the question as to whether Vodafone or Telefónica would be prepared to invest on an equivalent scale were Ofcom to authorise a launch using 900 MHz spectrum (where, admittedly, the only compatible handsets were the Galaxy SIII and Lumia 820) (Gabriel 2012c).

The fact that the EE brand was being used specifically for the LTE network effectively suggested that it was a premium brand compared to the Orange and T-Mobile brands still being used by its parents (albeit with the distinct possibility that these would be phased out in due course). As such, it could just about be seen as a new competitor in the market rather than as an incumbent with an added advantage. On that basis, all incumbents agreed to take no action in the courts before mid-October pending any further decisions by Ofcom concerning use of the 900 MHz band for 4G (TeleGeography 2012o). In early October, Ofcom sought to forestall legal action by indicating that it would bring forward the auction of digital dividend spectrum, as detailed below, in order to make it possible for EE's rivals to launch by May 2013 (Grierson 2012).

At the time of the launch, EE claimed that it would be able to provide a 12–15 Mbps downlink. Its basic service would provide 500 MB of data for £36 a month, while a more realistic 8 GB would cost as much as £56. Rivals immediately cast doubt on the ability of an 1,800 MHz network to supply a strong signal indoors, and Vodafone offered a 70 % reduction on the cost of an outstanding contract to any customer willing to sign on for its own LTE service once it was launched.

For its part, UK Broadband (owned by Hong Kong's PCCW) announced in October 2012 the commercial launch of a TD-LTE network, albeit restricted initially to Southwark in London and Reading (Cellular-news 2012d). This was claimed to be the world's first TD-LTE network in the 3.5/3.6 GHz bands in which it holds 124 MHz of spectrum, allowing for six 20 MHz channels to be utilised. As the holder of the only national WiMAX licence in the UK, it intends to operate as a wholesaler (UK Broadband 2012).

In June 2012, Vodafone and Telefónica announced that they had signed a 50/50 infrastructure-sharing agreement initially covering their 2G and 3G networks. However, they would be bidding independently in any subsequent spectrum auctions (Telecom.paper 2012g). The plan, creating Cornerstone Telecommu-nications Infrastructure and promising 98 % population coverage across all technologies within 3 years, was approved by the Office of Fair Trading in October (Wood 2012).

Details of the 800 MHz and 2.6 GHz auctions, to be held simultaneously using a combinatorial clock format and providing indefinite duration, technology-neutral licences, finally emerged in July 2012 (Ofcom 2012). Ofcom began by reiterating that it wanted to preserve 'at least four credible national wholesalers of mobile services', a requirement that could only be met if either H3G or a new entrant won sufficient spectrum in the combined auction. Hence, a licence would have to be set aside for a bidder other than Everything Everywhere, Telefónica and Vodafone. Provided at least one such bidder was willing to pay the reserve price, the highest offer would be accepted. In order to maintain a reasonably level playing field, safeguard caps on spectrum holdings would apply as follows: an overall cap of 105 MHz paired and a sub-1 GHz cap of 27.5 MHz paired—in both cases covering all spectrum won in the auction together with all of an operator's spectrum holdings in the 900 MHz, 1,800 MHz and 2.1 GHz paired bands (but excluding 2.1 GHz band unpaired spectrum for which no commercial use yet existed).

Rather than specify one particular optimum spectrum portfolio for the fourth national wholesaler, Ofcom detailed five possible portfolios, of which two included 1,800 MHz spectrum. It went on to state that no 2.6 GHz spectrum would be reserved for low-power shared use by operators of sub-national Radio Access Networks. On the matter of coverage obligations, Ofcom restricted these to a single 800 MHz licence providing 10 MHz paired, while expressing the opinion that the forces of competition would drive other licensees to follow suit. The primary requirement, to be met by 31 December 2017, was that this licensee should provide indoor coverage to 98 % of the UK population (expected to be equivalent to outdoor population coverage of 99.5 %), with minimum indoor coverage set at 95 % for each individual country within the UK.

With respect to the 800 MHz band, aside from the 10 MHz paired set aside for the licence with coverage obligations (at a reserve price of £250 million), all of the spectrum would be offered in blocks of 5 MHz paired (at a reserve price of £225 million). With respect to the 2.6 GHz band, the spectrum would be offered in blocks of 5 MHz paired and 5 MHz unpaired.

In the light of the above discussion concerning use of the 1,800 MHz band for LTE, it is evident that—assuming the acquisition of 800 MHz and 2.6 GHz spectrum by Vodafone and Telefónica in the auction—the gap between the launch by EE and that by its main rivals will be reduced to approximately 6 months, which should not disadvantage them unduly although it remains the case that 1,800 MHz has become the preferred band for LTE in Europe. However, they will have to compete on price, which may or may not prove to be problematic—for details of EE's initial tariffs published in October 2012 see Tomaszczyk (2012). It is possible to argue that H3G is a special case because it has ploughed ahead with upgrading its existing network to DC-HSPA+, branded as 'Ultrafast', achieving 40 % population coverage (located in large cities) by November 2012. The difference in the downlink between that and EE's LTE service may not prove to be all that great in practice (Gabriel 2012e). This may partly account for why EE increased its data allowances by 60 % in November, but it was also the case that EE was accused of taking advantage of its temporary monopoly by providing unusually small data allowances at each price point.

In October 2012, the four incumbents formed a jointly-owned company called Digital Mobile Spectrum Ltd (DMSL) in order to facilitate the rapid spread of LTE in the 800 MHz band. In particular, steps would be taken to ensure that there would be no interference with Freeview TV signals. The cost would be borne by all successful bidders for spectrum in the band (TeleGeography 2012q).

In November, Ofcom announced that it was working on plans to make available the 700 MHz band—currently used for digital terrestrial TV (DTT) and hence not requiring an expensive switchover from analogue. It referred to this as 'possibly

5G', which can only best be described as unhelpful except in the sense that it is not going to come on stream until 2018 at the earliest by which time 'true 4G' will have become available. The band is already in use elsewhere in the world, especially in the USA, and is expected to be sanctioned for world- wide use by the World Radio Conference in 2015. However, that is only a first step in the long process of securing international harmonisation in the band (Lennighan 2012c).

At the end of December there were three developments. Firstly, Ofcom announced that in addition to the four incumbents, HKT (UK) (a subsidiary of Hong Kong's PCCW), Niche Spectrum Ventures (a subsidiary of fixed-wire incumbent BT) and managed networks supplier MLL Telecom had registered as potential bidders in the forthcoming auction (TeleGeography 2012u). However, BT—already a mobile MVNO—had only expressed an interest in the 2.6 GHz band as it was conducting trials using white spaces in the TV spectrum for potential use in the provision of rural broadband.

Secondly, the Ministry of Defence stated that it would be auctioning off 200 MHz of its spectrum below 15 GHz with a target date for completion of summer 2014 (Sahota 2012). Thirdly, EE announced that LTE would be available in 18 cities and also that it would be launching in Northern Ireland by the year-end. Recent tests had indicated an average downlink of 17 Mbps outdoors and 9.7 Mbps indoors.

2013 opened predictably with further controversy when EE announced, firstly, that it had signed up its first MVNO, major retailer Phones4U (although no immediate launch was pencilled in) and, secondly, a new set of LTE tariffs. While this included a 20 GB per month plan for the heaviest users at a commensurately high price, the main objection was to the special offers available until the end of March for more modest consumers who would be expected to lock themselves in for 2 years before any rivals had the opportunity to publish their own tariffs. In early February, H3G struck back by announcing that it would not be charging a premium for its LTE service while omitting to specify when this would become available. As was duly noted, it seemed unreasonable to charge a premium for something that almost no-one could access, although H3G could still cause surprise with a swift roll-out and it would be receiving the first instalment of 1,800 MHz spectrum from EE before end-September.

At the end of January 2013, Vodafone announced a voice/data plan for the BlackBerry Z10 which had been launched the previous day, the first plan available for a '4G-ready' handset, while EE stated that it had achieved LTE coverage of 45 % of the population.

For its part, Ofcom proposed to make licences in the 900 MHz, 1,800 MHz and 3G bands technology-neutral. While this appeared to be very forward-looking, the response was muted because there was little expectation that operators would want to cannibalize their 900 MHz and 3G networks until they were ready to switch all customers onto LTE, especially given the paucity of compatible devices in these bands. In any event, experience had already shown that certain operators were wholly opposed to the possibility of 900 MHz re-farming being used as an excuse for reallocations of spectrum in the band (Gabriel 2013a).

Table 4.1 Auction results 2013

	800 MHz	2.6 GHz	£ million
H3G	5 paired	–	225.0
Vodafone	10 paired	20 paired, 25 unpaired	790.7
Telefónica	10 paired	–	550.0
EE	5 paired	35 paired	588.9
Niche Spectrum	–	15 paired, 20 unpaired	184.5
Total	60	185	2,341.1

Prior to the auction, existing holdings were as follows:

- Vodafone—17.5 MHz paired/900 MHz; 6 MHz paired/1,800 MHz; 15 MHz paired/3G. Total = 38.5 MHz paired.
- Telefónica—17.5 MHz paired/900 MHz; 6 MHz paired/1,800 MHz; 10 MHz paired/3G. Total = 33.5 MHz paired.
- H3G—0 MHz paired/900 MHz; 0 MHz paired/1,800 MHz; 15 MHz paired/3G. Total = 15 MHz paired.
- EE—0 MHz paired/900 MHz; 45 MHz paired/1,800 MHz; 20 MHz paired/3G. Total = 65 MHz paired.

However, EE was committed to the sale to H3G of 15 MHz paired of 1,800 MHz spectrum. This meant that, subject to the overall cap of 105 MHz paired and the sub-1 GHz cap of 27.5 MHz paired and pending the transfer of spectrum between EE and H3G:

- Vodafone could acquire up to 66.5 MHz paired overall of which no more than 10 MHz paired could be sub-1 GHz.
- Telefónica could acquire up to 71.5 MHz paired overall of which no more than 10 MHz paired could be sub-1 GHz.
- EE could acquire up to 40 MHz paired overall of which no more than 27.5 MHz paired could be sub-1 GHz.
- H3G could acquire up to 90 MHz paired overall of which no more than 27.5 MHz paired could be sub-1 GHz.

The results of the auction which terminated in mid-February were as follows: Table 4.1.

These results raised a number of interesting points. In the first place, the amount raised fell well short of the £3.5 billion that had been included in the government's Autumn Statement (Gabriel 2013b) and this triggered an investigation by the National Audit Office. Secondly, Telefónica somewhat surprisingly ended up with only the single block of 800 MHz spectrum to which the coverage obligations set out previously were attached. It will have to use this for rural coverage as in Germany, which begs the question as to how it intends to achieve full coverage in urban areas. It has been suggested that it will either have to re-farm its 3G spectrum or act as a MVNO on UK Broadband's 3.5 GHz network. Thirdly, Niche Spectrum did well to acquire both paired and unpaired spectrum in the 2.6 GHz band. This was also the case for Vodafone so both Vodafone and BT will be joining the so-far limited ranks of operators able to provide simultaneously both FDD and TDD versions of LTE (Newlands 2013). Finally, it may be noted that only Vodafone now conforms to the most typical European set-up of a joint 800 MHz/2.6 GHz network although EE will be able to use a combined 800 MHz/1,800 MHz/2.6 GHz network as against the less favoured combination of 800 MHz/1,800 MHz to be launched by H3G. However, Vodafone decided to delay its launch, provisionally intended for June 2013, until a later date as the current model of the iPhone was incompatible with the spectrum it intended to use.

In February 2013, the auction winners were formally allocated their licences. BT paid £15.1 million in order to be given the 2,520–2,535 MHz band paired with the 2,640–2,655 MHz band while Vodafone paid £8.1 million for 801–811 MHz paired with 842–852 MHz plus £4 million for 2,500–2,520 MHz paired with 2,620–2,640 MHz. Telefónica automatically received without payment 811–821 MHz paired with 852–862 MHz. These sums made negligible difference to the total amount raised. Ofcom stated in July that all 800 MHz spectrum was now cleared of previous users.

In March, EE stated that its LTE network now covered 50 % of the population It added that 80 towns and cities would be covered by the end of June and 70 % of the population by the end of the year (TeleGeography 2013j). It also revealed plans to trial Carrier Aggregation while brushing aside any suggestions that it was rushing to launch LTE-A because of the lack of Category 4 devices. However, by using another 20 MHz carrier in the 1,800 MHz band it expected to double the download speed to a maximum of 80 Mbps, commencing in 10 major cities later in the year (Wood 2013a). It claimed to have 318,000 LTE subscribers at the end of April and over 500,000 by the end of May. However, the 800 MHz spectrum is problematic because it strictly needs to be used in carriers of 10 MHz paired, and as a result EE and H3G began negotiations in May 2013 with a view to combining their respective 5 MHz paired holdings. An agreement between these two parties is a little surprising given that H3G has promised not to charge a premium for its 4G services which has certainly not been EE's preferred strategy so far.

In July, EE stated that it had turned on double-speed LTE in 12 cities, thereby raising the theoretical maximum downlink to 150 Mbps and doubling the average downlink to 48–60 Mbps from 24–30 Mbps. A range of much improved subscriber plans for those signing 2-year contracts was also announced (Telecom.paper 2013e).

Also in July, Ofcom announced that it would be proceeding with variations in licences for the 900 MHz, 1,800 MHz and 2,100 MHz bands in order to align them more closely with conditions pertaining to newer licences and to permit re-farming for LTE and WiMax.

There is considerable debate concerning how BT will proceed (Wood 2013b). It has signed up with Telefónica to provide transmission and backhaul infrastructure over a 10-year period but the agreement makes no mention of mobile spectrum. It is felt that Telefónica would benefit from leasing part of BT's 2.6 GHz blocks—BT has excluded the possibility of building its own network.

References

BBC. (2012). *Everything Everywhere gets 4G go-ahead from Ofcom.* Accessed August 21, 2012, from http://www.bbc.co.uk

Cauchi, M. (2012). *Vodafone UK chief slams Ofcom over Everything Everywhere.* Accessed March 19, 2012, from http://www.totaltele.com

Cellular-news. (2010). *Qualcomm to sell Indian BWA spectrum?* Accessed June 11, 2010, from http://www.cellular-news.com

Cellular-news. (2011a). *UK may delay 4G radio spectrum auction.* Accessed September 7, 2011, from http://www.cellular-news.com

Cellular-news. (2011b). *BSNL offers to hand back BWA spectrum – wants refund on license fee.* Accessed October 19, 2011, from http://www.cellular-news.com

Cellular-news. (2012a). *UK regulator issues another consultation on its 4G license auction.* Accessed January 16, 2012, from http://www.cellular-news.com

Cellular-news. (2012b). *Hutchison could sell UK network if 4G license terms are not favourable.* Accessed May 22, 2012, from http://www.cellular-news.com

Cellular-news. (2012c). *MTS launches NSN supplied LTE network in Moscow.* Accessed September 19, 2012, from http://www.cellular-news.com

Cellular-news. (2012d). *UK based ISP launches fixed wireless LTE service.* Accessed October 8, 2012, from http://www.cellular-news.com

Cellular-news. (2012e). *India to lower reserve prices on unsold GSM spectrum.* Accessed December 10, 2012, from http://www.cellular-news.com

Gabriel, C. (2010). *Infotel emerges as big winner in Indian BWA auction.* Accessed June 13, 2010, from http://www.rethink-wireless.com

Gabriel, C. (2012a). *UK could see 1.8GHz auction within a month.* Accessed January 11, 2012, from http://www.rethink-wireless.com

Gabriel, C. (2012b). *Ofcom removes spectrum guarantees for two UK operators.* Accessed January 13, 2012, from http://www.rethink-wireless.com

Gabriel, C. (2012c). *Would Vodafone and O₂ refarm 900MHz spectrum in UK?* Accessed September 16, 2012, from http://www.rethink-wireless.com

Gabriel, C. (2012d). Virgin Media backs away from UK auctions. Accessed September 25, 2012, from http://www.rethink-wireless.com

Gabriel, C. (2012e). *3UK to bring DC-HSPA+ to 80% of UK by April.* Accessed November 22, 2012, from http://www.rethink-wireless.com

Gabriel, C. (2013a). *UK plans to free up all mobile bands for LTE.* Accessed February 11, 2013, from http://www.rethink-wireless.com

Gabriel, C. (2013b). *UK operators may need to trade 4G spectrum.* Accessed March 18, 2013, from http://www.rethink-wireless.com

Government of India. (2010, February 25). *Auction of 3G and BWA spectrum: Notice inviting applications.* File No: P-11014/13/2008-PP.

Grierson, J. (2012). *Mobile 4G crunch talks to be held.* Accessed October 2, 2012, from http://www.themediaguru.blogspot.com

Grundberg, S. (2012). *Tele2 CEO sees no chance of getting 3G licence in Russia.* Accessed February 7, 2012, from http://www.totaltele.com

House of Commons Culture, Media and Sport Committee. (2011). *Spectrum.* Eighth report of session 2010–12, November. London: The Stationery Office.

Kinetz, E. (2010). *Reliance Ind. Pays $1B for India broadband winner.* Accessed June 15, 2010, from http://www.cellular-news.com

Krishna, R. (2012a). *India starts telecom spectrum auction.* Accessed November 12, 2012, from http://www.cellular-news.com

Krishna, R. (2012b). *India spectrum auction generates $1.71 billion.* Accessed November 14, 2012, from http://www.cellular-news.com

Lennighan, M. (2012a). *Friday review: India in disarray.* Accessed February 6, 2012, from http://www.totaltele.com

Lennighan, M. (2012b). *UK gets LTE with EE and BJ.* Accessed September 11, 2012, from http://www.totaltele.com

Lennighan, M. (2012c). *UK starts planning for 5G mobile.* Accessed November 19, 2012, from http://www.totaltele.com

Lennighan, M. (2013). *Russian telcos angered by threat of LTE spectrum seizure.* Accessed July 8, 2013, from http://www.totaltele.com

Middleton, J. (2011). *Yota scores deal as Russia's LTE network operator.* Accessed March 3, 2011, from http://www.telecoms.com

Newlands, M. (2013). *iPhone 5 will initially support LTE bands in nine countries.* Accessed January 20, 2013, from http://www.cellular-news.com

Ofcom. (2011). *Consultation on assessment of future mobile competition and proposals for the award of 800 MHz and 2.6 MHz spectrum and related issues.* London: Ofcom.

Ofcom. (2012). *Assessment of future mobile competition and award of 800 MHz and 2.6 GHz. Statement.* London: Ofcom.

Sahota, D. (2012). *MoD to auction off spectrum for LTE.* Accessed December 19, 2012, from http://www.telecoms.com

Smith, C. (2012). *Three clarifies '4G' claims, while first LTE network hits London.* Accessed March 2, 2012, from http://technowizz.wordpress.com

Telecom Regulatory Authority of India. (2012). *Consultation paper on auction of spectrum.* New Delhi: TRAI.

Telecom.paper. (2012a). *India to allocate 700MHz band for 4G services.* Accessed March 7, 2012, from http://www.telecompaper.com

Telecom.paper. (2012b). *Airtel to launch LTE-TDD network this month.* Accessed April 4, 2012, from http://www.telecompaper.com

Telecom.paper. (2012c). *Rostelecom, Yota sign agreement on LTE access.* Accessed April 17, 2012, from http://www.telecompaper.com

Telecom.paper. (2012d). *Yota to merge with Megafon holding company.* Accessed April 27, 2012, from http://www.telecompaper.com

Telecom.paper. (2012e). *Qualcomm receives Indian licence with reduced term.* Accessed May 11, 2012, from http://www.telecompaper.com

Telecom.paper. (2012f). *Airtel acquires 49% stake in Qualcomm BWA business.* Accessed May 24, 2012, from http://www.telecompaper.com

Telecom.paper. (2012g). *Vodafone, O_2 to set up UK network-sharing joint venture.* Accessed June 7, 2012, from http://www.telecompaper.com

Telecom.paper. (2012h). *Summa Telecom wins court claim on LTE frequencies.* Accessed July 16, 2012, from http://www.telecompaper.com

Telecom.paper. (2012i). *India to allow operators to share spectrum access.* Accessed October 25, 2012, from http://www.telecompaper.com

Telecom.paper. (2012j). *Idea emerges as aggressive bidder for 2G auction.* Accessed October 30, 2012, from http://www.telecompaper.com

Telecom.paper. (2012k). *India may allow GSM operators to keep 900MHz spectrum.* Accessed November 1, 2012, from http://www.telecompaper.com

Telecom.paper. (2012l). *India GSM auction falls short of expectations.* Accessed November 15, 2012, from http://www.telecompaper.com

Telecom.paper. (2012m). *Beeline secures LTE access on Yota network.* Accessed December 10, 2012, from http://www.telecompaper.com

Telecom.paper. (2013a). *India allows BWA spectrum holders to offer voice services.* Accessed February 21, 2013, form http://www.telecompaper.com

Telecom.paper. (2013b). *Indian spectrum auction uncertain after just 1 applicant.* Accessed February 2013, from http://www.telecompaper.com

Telecom.paper. (2013c). *Yota spectrum licence overturned in interference dispute.* Accessed March 5, 2013, from http://www.telecompaper.com

Telecom.paper. (2013d). *Yota faces competition investigation over LTE access.* Accessed March 12, 2013, from http://www.telecompaper.com

Telecom.paper. (2013e). *M1, SingTel, StarHub secure 4G spectrum.* Accessed July 1, 2013, from http://www.telecompaper.com

TeleGeography. (2011a). *BSNL, MTNL join Aircel in increasing readiness for 4G LTE.* Accessed August 17, 2011, from http://www.telegeography.com

TeleGeography. (2011b). *MTS protests against loss of WiMAX subscribers following new LTE scheme.* Accessed September 13, 2011, from http://www.telegeography.com

TeleGeography. (2011c). *Yota requests Moscow LTE examination; hints at new MegaFon tie-up.* Accessed November 30, 2011, from http://www.telegeography.com

TeleGeography. (2011d). *Rostelecom given belated clearance for LTE rollout.* Accessed December 2, 2011, form http://www.telegeography.com

TeleGeography. (2012a). *MTS awarded Moscow LTE licence by Roskomnadzor.* Accessed February 2, 2012, from http://www.telegeography.com

TeleGeography. (2012b). *Osnova sounds LTE battle-cry; is MoD-backed start-up now armed with a 4G licence?* Accessed March 7, 2012, from http://www.telegeography.com

TeleGeography. (2012c). *WiMAX operator Enforta poised to test LTE in Kemerovo?* Accessed March 13, 2012, from http://www.telegeography.com

TeleGeography. (2012d). *Antares authorised to use 1900MHz band for LTE.* Accessed May 1, 2012, from http://www.telegeography.com

TeleGeography. (2012e). *MegaFon plots TD-LTE network to complement Yota MVNO.* Accessed May 5, 2012, from http://www.telegeography.com

TeleGeography. (2012f). *Roskomnadzor announces July tender for LTE frequencies in 791MHz-862MHz band.* Accessed May 5, 2012, from http://www.telegeography.com

TeleGeography. (2012g). *Government considering regional LTE tenders.* Accessed May 16, 2012, from http://www.telegeography.com

TeleGeography. (2012h). *Minnow Summa set to join mobile giants in LTE tender; Tele2 unhappy with terms of regional auction.* Accessed June 8, 2012, from http://www.telegeography.com

TeleGeography. (2012i). *MegaFon chalks up 7,000 LTE modem sales in Moscow in May.* Accessed June 13, 2012, from http://www.telegeography.com

TeleGeography. (2012j). *Yota reveals details of LTE transition.* Accessed June 15, 2012, from http://www.telegeography.com

TeleGeography. (2012k). *MTS to launch LTE in Moscow on 1 September.* Accessed July 31, 2012, from http://www.telegeography.com

TeleGeography. (2012l). *Rostelecom pinpoints LTE launch in North Caucasus in 2013.* Accessed August 20, 2012, from http://www.telegeography.com

TeleGeography. (2012m). *Chaos in the Kosmos as pay-TV firm shuts down after LTE spectrum compensation.* Accessed August 21, 2012, from http://www.telegeography.com

TeleGeography. (2012n). *See you in court! Rostelecom initiates legal action against Scartel over LTE access.* Accessed September 11, 2012, from http://www.telegeography.com

TeleGeography. (2012o). *UK operators agree not to launch legal action over 4G rollout plans for a month.* Accessed September 11, 2012, from http://www.telegeography.com

TeleGeography. (2012p). *DoT postpones refarming decision, allows operators to swap spectrum.* Accessed September 26, 2012, from http://www.telegeography.com

TeleGeography. (2012q). *Operators form joint venture to speed up 800MHz 4G deployment.* Accessed October 19, 2012, form http://www.telegeography.com

TeleGeography. (2012r). *India spectrum auction falls short of expectations, SSTL remains on sidelines.* Accessed October 24, 2012, from http://www.telegeography.com

TeleGeography. (2012s). *Aircel ditches 4G until next year.* Accessed October 26, 2012, from http://www.telegeography.com

TeleGeography. (2012t). *No new auction this fiscal year says TRAI.* Accessed December 3, 2012, from http://www.telegeography.com

TeleGeography. (2012u). *Ofcom unveils identities of 4G bidders.* Accessed December 21, 2012, from http://www.telegeography.com

TeleGeography. (2013a). Federal arbitration court makes valentine's day date with summa over 4G fallout. Accessed January 10, 2013, from http://www.telegeography.com

TeleGeography. (2013b). *Mobile giant makes a Beeline towards Moscow LTE launch.* Accessed January 15, 2013, from http://www.telegeography.com

TeleGeography. (2013c). *Govt sets out auction plans.* Accessed January 23, 2013, from http://www.telegeography.com

TeleGeography. (2013d). *New defence minister offers to return Osnova's LTE frequencies.* Accessed February 11, 2013, from http://www.telegeography.com

TeleGeography. (2013e). *Russia round-up: Osnova owner disputes MoD frequency plans; Antares LTE hearing postponed.* Accessed February 12, 2013, from http://www.telegeography.com

TeleGeography. (2013f). *Vainakh sees regional ruckus over Siberian LTE equipment.* Accessed March 1, 2013, from http://www.telegeography.com

TeleGeography. (2013g). *Moscow court chastises 'inactive' watchdog; finds in favour of Antares Telecom.* Accessed March 5, 2013, from http://www.telegeography.com

TeleGeography. (2013h). *Federal arbitration court rejects summa LTE appeal.* Accessed March 5, 2013, from http://www.telegeography.com

TeleGeography. (2013i). *Telenor allowed to offset new spectrum fees.* Accessed March 7, 2013, from http://www.telegeography.com

TeleGeography. (2013j). *EE LTE rollout reaches 50-city milestone five months after launch.* Accessed March 28, 2013, from http://www.telegeography.com

TeleGeography. (2013k). *Vimpelcom launches LTE network in Moscow.* Accessed May 28, 2013, from http://www.telegeography.com

TeleGeography. (2013l). *Rostelecom launches LTE in Sochi.* Accessed June 4, 2013, from http://www.telegeography.com

Tomaszczyk, S. (2012). *EE unveil 4GEE tariffs.* Accessed October 23, 2012, from http://www.totaltele.com

UK Broadband. (2012). *UK Broadband switches on first commercial 4G TD LTE system in the UK.* Accessed July 9, 2012, from http://www.ukbroadband.com

Wood, N. (2012). *Competition watchdog approves Vodafone, Telefonica UK network-sharing plan.* Accessed October 2, 2012, from http://www.totaltele.com

Wood, N. (2013a). *EE bemoans lack of LTA-handsets.* Accessed March 9, 2013, from http://www.totaltele.com

Wood, N. (2013b). *O_2 UK strikes 10-year wholesale deal with BT.* Accessed April 29, 2013, from http://www.totaltele.com

Europe

5

5.1 Austria

In September 2010, 14 paired (FDD) and 10 unpaired (TDD) blocks, each of 5 MHz of spectrum in the 2.6 GHz band, were offered for sale by auction. Unusually, at least in respect of 2.6 GHz spectrum, the regulator required that 25 % of the population be provided with a downlink of 1 Mbps and an uplink of 256 kbps no later than end-December 2013, and this may partly account for why the auction raised only €39.5 million. A further factor was that a 'second price' rule was used whereby the winner pays the price offered by the runner-up. A1 Telekom Austria bought four paired and five unpaired blocks and announced that it would commence commercial operations in Vienna in December with its 'A1 Broadband LTE' service offering 30 GB at a theoretical downlink speed of up to 150 Mbps for €90 a month. Hutchison 3G Austria acquired four paired and five unpaired blocks, and pencilled in a launch for November 2011 (duly completed), while T-Mobile Austria bought four paired blocks and Orange bought two paired blocks. T-Mobile and Telekom Austria launched before the end of 2010 although the former was restricted to selected corporate clients (soft launch) with a full commercial launch pencilled in for July 2011.

In early 2012, Hutchison (H3G) made a takeover bid for Orange. Pending the outcome of this, with the European Commission in November giving permission to proceed provided certain conditions designed to encourage competition were first implemented (TeleGeography 2012ab), the regulator put any further spectrum auctions on hold—it confirmed in October 2012 that no further auctions would take place before September 2013. It is of interest that T-Mobile's objection to the takeover was to the effect that if spectrum was to be redistributed between H3G and Telekom Austria and multiple antenna sites sold on to Telekom Austria, the two operators would be able to speed up their LTE network roll-outs to the detriment of competitors—while omitting to note that T-Mobile had long held the spectrum bank best suited for re-farming for LTE (TeleGeography 2012s). The regulator authorised spectrum transfers from Orange to H3G and from H3G to Telekom

P. Curwen and J. Whalley, *Fourth Generation Mobile Communication,*
Management for Professionals, DOI 10.1007/978-3-319-02210-9_5,
© Springer International Publishing Switzerland 2013

Austria in December 2012, and T-Mobile immediately appealed the decision but without initial success. However, in May 2013 the Vienna Administrative Court referred the issue to the European Court of Justice (ECJ) (TeleGeography 2013h).

Meanwhile, in September 2012, Telekom Austria—which had been the first operator to introduce a 4G-enabled handset in March—claimed that its network now covered all provincial capitals and major conurbations, making it the biggest in Austria. One-quarter of the population would have access to LTE by the year-end.

It was reported in December that T-Mobile, in partnership with Huawei, had begun testing LTE-Advanced in Vienna using a combination of spectrum in the 1,800 MHz and 2.6 GHz bands, in the process achieving a downlink of 289 Mbps (TeleGeography 2012ac).

In March 2013, the regulator confirmed that, with the Hutchison take-over of Orange now completed (in January with the Orange brand to be phased out by August), it would be auctioning 28 blocks, each of 5 MHz paired, in the 800 MHz, 900 MHz and 1,800 MHz bands in September with the aggregate reserve price set at €526 million ($680 million). 10 MHz paired in the 800 MHz band, valued at a minimum of €45.6 million, would be reserved for a new entrant that would also be able to secure roaming rights. However, it was noted that, post-merger, Hutchison would own 29 MHz paired in the 1,800 MHz band and 4 MHz paired in the 900 MHz band, thereby consolidating its position as the third incumbent and leaving little opportunity for a new entrant to establish a foothold in the market, especially given Hutchison's ability to speed up its LTE roll-out in the 2.6 GHz band. The auction method chosen was so complicated that T-Mobile even hired a theoretician to explain how it worked (TeleGeography 2013g).

Whether the auction can now take place in September is a moot point given T-Mobile's appeal before the ECJ. The regulator had already recognised the implications of the delay for the re-farming of 2G/3G spectrum for LTE (in October 2012) which was supposed to take place after the auction, and hence this is going to proceed during 2013 irrespective of when the auction occurs.

A tender for spectrum in the 450 MHz band was issued in March, consisting of 21 blocks, each of 200 kHz paired at a reserve price of €17,000 and licensed for 15 years (lasting until end-2029). This spectrum had originally been used for analogue transmissions and, when these ceased, subsequently assigned to T-Mobile and Sweden-based Green Network for €6 million by way of an auction in 2006. After the spectrum was returned in 2008, a second auction took place but this attracted no applicants. While this band is technically a competitor with digital dividend spectrum for rural coverage it has yet to prove attractive to incumbents, but the regulator believes that there will be real interest expressed in financial terms this time around.

Telekom Austria is pushing ahead with trials of LTE-Advanced. It claimed to have achieved a 580 Mbps downlink in June 2013.

5.2   Belgium

It was determined that 15-year licences covering 185 MHz (70 MHz paired FDD + 45 MHz TDD) in the 2.6 GHz band would be auctioned at the end of November 2011 and become operational in July 2012. Incumbents, which were conducting trials, would be capped at 20 MHz paired. In October, five applications were received from incumbents Belgacom, Mobistar (Orange) and KPN together with Craig Wireless and BUCD. The winners were Belgacom (20 MHz paired at a cost of €20.2 million), BUCD (45 MHz TDD at a cost of €22.5 million), KPN (15 MHz paired at a cost of €15.0 million) and Orange (20 MHz paired at a cost of €20.0 million). Craig Wireless was unsuccessful due to non-participation in the bidding. In total, €77.8 million ($103.7 million) was raised (Rayal 2011) but 15 MHz paired was left unsold, essentially due to the presence of only three incumbents. Most unusually, the unpaired central block fetched a higher price than the surrounding paired blocks. It is thought that BUCD is Asian-owned, possibly by China Mobile.

In June 2012, Belgacom announced that it was busy rolling out LTE over its 1,800 MHz network and that it hoped to launch before the end of 2012. Eight areas would initially be covered but for technical reasons relating to radiation levels these would not include Brussels (see below). The 2.6 GHz spectrum would be utilised at a later date as required. The launch duly took place in November, providing a service over a variety of devices (Telecom.paper 2012i). For its part, Orange pencilled in 2013 for its launch.

Meanwhile, in May, Clearwire Belgium, now owned by TechMax, announced that it intended to launch a nationwide TD-LTE network before the year-end using its 3.5 GHz (non-mobile) licence. This was subsequently amended to a launch in Brussels using the brand 'b-lite' (Wood 2012).

Environmental restrictions are not uncommon, but they are particularly vexing in Brussels. The EU and World Health Organisation issue recommendations concerning matters such as radiation limits at base stations, but there is nothing to stop local authorities from imposing stricter limits and Brussels has fixed a 3 volt per metre (V/m) maximum at base stations which is allegedly 200 times below the recommended level. By comparison, for example, Walloon and Flanders in general allow 3 V/m per antenna and per operator which is much more generous, but the situation is set to worsen when tighter restrictions are imposed in Brussels in 2015. The irony, of course, is that Brussels is the headquarters of the European Commission which is unlikely to benefit from LTE any time soon other than that provided by b-lite which, having no 2G or 3G network, can meet the radiation limit without difficulty.

The 800 MHz band is to be sold via a SMRA auction in early November in the form of three licences, each providing 10 MHz paired at a minimum price of €120 million to three independent licensees. Licences won by incumbents will come with a population coverage requirement of 98 % within 6 years of licence receipt whereas a new entrant will be given a further 3 years to reach that target. In addition, whereas a successful incumbent will be obliged to offer national roaming, a successful new entrant will be able to demand it. The 15 MHz paired of 2.6 GHz spectrum left over from 2011 is to be reserved for a successful bidder that has yet to acquire spectrum in that band (TeleGeography 2013m).

5.3   Bulgaria

In December 2011, Max Telecom and 4G Com won 8 MHz paired in the 1,800 band while Bulsatcom won 5 MHz paired. However, it was claimed that only Bulsatcom had paid for its 10-year licence (at a cost of $13 million) which was claimed to be restricted to WiMAX, LTE or LTE-Advanced although the regulator subsequently referred to it as the fourth GSM licence, and it was the only one to be handed a licence in January 2013 (TeleGeography 2013b). Despite this, Max Telecom, which had offered $19 million, announced in June that it would be launching LTE in 2013Q4 so it had evidently received its licence in the meantime.

For their part, the incumbent mobile operators began to conduct tests using the 1,800 MHz band. Telekom Austria appears to have launched a corporate service in January 2012 (4G Americas 2013) but it is debatable whether this should be considered as a commercial launch.

In April 2013, the government announced that the military would be commencing its departure from the 2.6 GHz band on 1 January 2014. The regulator tabled a plan for the use of the vacated spectrum in July 2013 which specified the sale of five 5 MHz paired blocks in the 2,500–2,570 MHz band paired with the 2,620–2,690 MHz band and three 5 MHz blocks in the 2,570–2,620 MHz band (TeleGeography 2013p). The 10-year licences would come with an obligation to launch 3G and/or 4G covering 35 % of the country within 2 years of licence issue and 55 % within 5 years of licence issue.

5.4   Croatia

In March 2012, T-Mobile (T-Hrvatski Telekom) claimed to have launched in four cities offering a (highly theoretical) maximum downlink of 173 Mbps and uplink of 58 Mbps. For its part, Telekom Austria (VIPnet) implemented a trial using 800 MHz spectrum which was completed in April 2011 and was followed by further trials using the 1,800 MHz and 2.6 GHz bands. It went on to claim that it had launched on the same day as T-Mobile, leading to the usual dispute as to which operator had been the first to launch (TeleGeography 2012g)—subsequent reports gave the launch date for T-Mobile as late April. The 1,800 MHz band appears to have been used for both launches, but it is questionable whether there were any devices or tariffs on offer in March.

30 MHz paired of digital dividend spectrum, freed up in October 2010, was put up for auction to existing operators at the end of October 2012. Three 12-year licences, each providing 10 MHz paired, were put on offer for a minimum one-off fee of $12.5 million. Licensees were obliged to achieve a minimum of 50 % land coverage within 5 years of the license issue date. However, the regulator noted that

there were issues arising from the staggered introduction of this spectrum in neighbouring countries that would affect potential coverage. In the event, only T-Mobile and Telekom Austria entered bids, with each offering €20 million ($26 million). T-Mobile is using the spectrum for rural coverage, launching with a maximum downlink of 75 Mbps in December 2012 (Telecom.paper 2012l).

Velatel is deploying TD-LTE using its holding of 42 MHz of spectrum in the 3.5 GHz band.

5.5   Cyprus South

There are two incumbents, CyTA and MTN. In June 2013, the government offered for sale by auction 10 MHz paired in the 900 MHz band (880–890 MHz paired with 925–935 MHz), 24.8 MHz paired in the 1,800 MHz band (1,760.2–1,785 MHz paired with 1,855.2–1,880 MHz) and 15 MHz paired in the 2,100 MHz band (1,950–1,965 MHz paired with 2,140–2,155 MHz). The spectrum was offered on a technology-neutral basis at a minimum price of €12.3 million with a view to attracting a new entrant into the market. The successful bidder will be required to achieve 40 % geographical coverage within 3 years of receipt of a licence and 65 % within 5 years. A minimum downlink of 30 Mbps will also be obligatory by 2019 (TeleGeography 2013l).

In July, the regulator announced an auction of spectrum in the 2.6 GHz band, comprising Category A—consisting of three blocks of 20 MHz paired (2,500–2,520 MHz paired with 2,620–2,640 MHz; 2,525–2,545 MHz paired 2,645–2,665 MHz; and 2,550–2,570 MHz paired with 2,670–2,690 MHz) at a minimum price of €2.7 million ($3.5 million) per block—and Category B— consisting of 40 MHz of TDD (2,575–2,615 MHz) at a minimum price of €2 million. Successful bidders will be obliged to cover 25 % of the country within 3 years and 40 % within 5 years from the date of licence receipt. Further, a minimum downlink of 30 Mbps will be obligatory by 2019 (TeleGeography 2013q).

A second attempt at a sale of spectrum in the 3.5 GHz band (3.4–3.8 GHz)— there was no interest expressed in 2011—may be the next step or possibly an auction of spectrum in the digital dividend band.

5.6   Czech Republic

In January 2012, the regulator announced that the winner of a multi-band auction— open to incumbents as well as new entrants who would be favoured—would be obliged to achieve 90 % population coverage within 3 years. In March, the date for an auction of 15-year licences in the 800 MHz, 1,800 MHz and 2.6 GHz bands was specified as November, with reserve prices amounting in total to €376 million ($496 million)—each 5 MHz paired/800 MHz at $56.5 million, each 5 MHz paired/ 2.6 GHz at $5.1 million and each 15 MHz paired/1,800 MHz at $23.1 million (Total Telecom 2012). A new entrant would obtain roaming rights on an incumbent's network if the latter bought at least 10 MHz paired in the 800 MHz band and held spectrum in the other 2G and 3G bands. A spectrum cap of 23 MHz paired in the 1,800 MHz band was to be applied as well as caps in the other bands.

In July, it was stated that 800 MHz spectrum would not now be offered as the regulator preferred to integrate the existing networks in this band via national roaming rights, but in practice all three bands were made available. The PPF Group, the wealthiest private equity group in central Europe, appeared to be the only viable bidder for 1,800 MHz spectrum, but there was considerable scepticism as to whether a new entrant could establish a profitable presence in the market (TeleGeography 2012n). The PPF Group (via PPF Mobile Services) duly applied whereas Dial Telecom, UPC, J&T, KKCG and Penta failed to do so. Licences were to be assigned no earlier than the end of 2012. The three incumbents plus PPF Mobile Services were accepted as bidders in October.

Meanwhile, in June 2012, Telefónica had launched in the small municipality of Jesenice using 10 MHz paired of re-farmed spectrum in the 1,800 MHz band (Sahota 2012). For its part, T-Mobile began a 3-month trial in Prague using Samsung's new 'Smart LTE' technology, and announced plans to build a LTE network in Mlada Boleslav before the year-end. A trial commenced there in November using 10 MHz paired in the 1,800 MHz band and providing a maximum 75 Mbps downlink for free over a 6-month period.

The auction was pencilled in for March 2013 and when it commenced a somewhat extraordinary event took place in that the regulator cancelled the auction because bids had risen to an unacceptably high level (in excess of $1 billion). As the regulator noted, the main purpose of the auction was to establish a new entrant and get LTE launched on the market within a short time frame, whereas bids had risen to a level which were incompatible with a swift roll-out of networks and the provision of services at an acceptable price level (Cellular-news 2013b). However, there were those who claimed that the auction design was flawed and that the incumbents had carried on bidding in order to deny the potential new entrant— sufficiently well-resourced to represent a real threat—access to an adequate block of spectrum at an economic price. Whatever the reason for the failure to proceed, it induced Telefónica to state that it now intended to continue to roll out LTE using re-farmed spectrum.

As yet there is no set date for a new auction and no-one appears to be happy with potential rule alterations such as reserving the 800 MHz block exclusively for a new entrant, although no specific objections have been tabled.

5.7 Denmark

The 2.6 GHz band licences were awarded in May 2010. The 20-year licences were technology-neutral. TeliaSonera won 20 MHz paired plus 10 MHz unpaired (as well as 15 MHz unpaired in the 2,010–2,025 MHz band) at a total cost of DKK336.3 million, Telenor won 20 MHz paired plus 10 MHz unpaired at a total cost of DKK333.3 million and TDC won 20 MHz paired at a total cost of

DKK333.3 million while Hutchison (Hi3G Access) won 10 MHz paired plus 25 MHz unpaired at a total cost of DKK7.1 million (Erhvervsstyrelsen 2010). If nothing else, this demonstrated the extremely low valuation placed upon unpaired spectrum at that point in time—see Ventura Team (2010)—although the paired spectrum sold for a relatively high price by European standards. TeliaSonera launched in December 2010 with coverage of 5 % of the population and TDC in October 2011.

As the only applicant, essentially because the three 2G incumbents were excluded from the bidding, Hi3G was also awarded 5 MHz paired in the 900 MHz band and 10 MHz paired in the 1,800 MHz band in September 2010 that could be used for LTE. It paid the minimum DKK8 million and DKK4 million respectively for the licences (TeleGeography 2010a).

In November 2011, TeliaSonera announced that its network was now available in 100 cities, with 60 % population coverage rising to 75 % by the year-end. It has an infrastructure-sharing agreement, TT-Netværket, with Telenor which includes 4G services, initially using the 1,800 MHz and 2.6 GHz bands. However, although TeliaSonera and Telenor have agreed to merge all of their networks, it was presumed that Telenor's official 4G launch would take place once the common 4G network was up-and-running (LteWorld 2012). In December 2012, Telenor postponed the date for its launch from January to March 2013 and promised 75 % population coverage.

30 MHz paired of digital dividend spectrum, divided into one lot of 10 MHz paired plus four lots of 5 MHz paired with a cap of 20 MHz paired per applicant, was awarded via 22-year nationwide licences in June 2012, with launches permitted after 1 January 2013. TDC won 20 MHz paired (801–821 MHz paired with 842–862 MHz)—the maximum permissible—at a cost of DKK628 million ($106 million) while TT-Netværket won 10 MHz paired (791–801 MHz paired with 832–842 MHz) at a cost of DKK114.5 million ($19.4 million)—20 % was payable up-front with a further 10 % annually (Telecom.paper 2012b). TT-Netværket will use this spectrum to extend its network into rural areas—the coverage requirement was expressed in terms of improving broadband access in the existing 'worst affected' areas. It may be noted that the improvement was expected to raise the downlink to an average of 10 Mbps, which was the fastest speed so far specified by a country in Europe, but in the event this only applied to TDC (Aetha 2012: p. 24).

In July, Hi3G announced that it intended to launch LTE in September with 37 % population coverage and that when the first 15 cities had full coverage by the year-end, this would rise to 50 % (TeleGeography 2012m). It would be using both the 1,800 MHz and 2.6 GHz bands at launch, and introducing the iPhone 5—for details of rate plans see Telecom.paper (2012e). It was also considering the use of unpaired spectrum for TD-LTE.

In February 2013, TeliaSonera reported that in the days immediately after it had been possible to connect the iPhone 5 to its LTE network for the first time, data traffic had surged by 40 %. It also noted that it had introduced international LTE roaming between Denmark and Sweden, claiming to be the first operator in Europe to provide such a service.

5.8   Estonia

In December 2010, 6 blocks of spectrum in the 2.6 GHz band were auctioned off, attracting 11 bids by 9 companies. The regulator awarded 20 MHz paired to TeliaSonera (which was also awarded 20 MHz (or possibly 40 MHz) unpaired) and to Elisa affiliates Saunalahti and Dial Media while Elisa subsidiary Elion was awarded 10 MHz paired. No financial details were forthcoming (TeleGeography 2010b). Tele2 announced that it would challenge the result as it had promised a quicker roll-out than Elisa, and the regulator responded by transferring the Dial Media spectrum—for which, it was alleged, it had refused to pay—to Tele2. TeliaSonera (EMT) launched in December 2010 and expected to achieve 70 % population coverage by the end of 2012. It noted that the proportion of customers with data plans signing up for LTE contracts had exceeded 50 % for the first time in November when the downlink provided between 20 Mbps and 60 Mbps (Cellular-news 2012j).

TeliaSonera and Tele2 already had permission to launch in the 1,800 MHz band, granted in December 2011, and TeliaSonera introduced this in conjunction with its 2.6 GHz network.

Tele2 launched in November 2012 in the northern regions of Estonia providing a theoretical maximum downlink of 150 Mbps. However, full nationwide coverage was not expected until 2017 (TeleGeography 2012z). Elisa was the third to launch in February 2013. So far, only TeliaSonera has provided handsets.

In May 2013, TeliaSonera won a beauty contest for a 10 MHz paired licence in the 800 MHz band—Tele2 and Elisa were the unsuccessful bidders This obliged it to install at least 199 base stations and cover 95 % of the country—which TeliaSonera achieved by end-June—in the process providing at least a 5 Mbps downlink. Two further licences are to be auctioned at an unspecified future date (Lennighan 2013).

5.9   Finland

Although it had used a beauty contest to assign 3G spectrum, Finland surprisingly resorted to an auction (necessitating new legislation) in the case of the 2.6 GHz band spectrum on offer in November 2009. The 20-year licences, providing 14 blocks of 5 MHz paired and 50 MHz of unpaired spectrum, raised a modest €3.8 million, much less than in Norway and Sweden. This was accounted for by the existence of only three incumbents, their extensive holdings of 2G and 3G spectrum, the fact that additional 1,800 MHz spectrum had been assigned in April 2009 in order to foster LTE, that the re-farming of GSM spectrum was permitted and that the auction format permitted switching behaviour. The licence winners were DNA, Elisa and TeliaSonera (four blocks, five blocks and five blocks of paired spectrum respectively)—the only bidders for paired spectrum—and Pirkanmaan Verkko, acting on behalf of the Finnet Group, which beat off one rival to acquire the unpaired spectrum with a view to rolling out WiMAX (Gabriel 2010). Because there was competition only for the unpaired spectrum, it sold for nearly twice as much per MHz per pop as the paired spectrum.

TeliaSonera initially launched a pre-commercial network in the centre of the Turku municipality in June 2010. On 30 November, a commercial launch, providing up to a theoretical maximum of 100 Mbps, took place in the Helsinki business centre and across the Turku municipality. The 'Sonera Full Net 4G' service cost €46 a month for 36 GB using a Samsung modem. The initial use of the 2.6 GHz band would give way to reliance upon the 1,800 MHz band which was first used for LTE in August 2011. For its part, Elisa launched in Espoo in December 2010 using both the 1,800 MHz and 2.6 GHz bands, while DNA launched in December 2011 while maintaining that its existing 1,800 MHz DC-HSPA+ service was also 4G.

Digital dividend spectrum, divided into six blocks of 5 MHz paired priced at a minimum €16.7 million per block, was originally to be auctioned before the end of 2013—for the background see Aetha (2011b)—but in April 2012 the government decided to start the process off during 2012 and to award the licences early in 2013. In October 2012, the regulator tabled FICORA Regulation 64 with a view to amending the existing Act covering the auction of 2.6 GHz spectrum so as to extend it to cover 800 MHz spectrum (TeleGeography 2012u). The licences would run for 20 years commencing 1 January 2014. No operator would be allowed to win more than three blocks. The coverage requirement would consist of 95 % of the population in 3 years and 97 or 99 % in 5 years.

In January 2013, the government announced its intention to amend legislation so as to permit the 700 MHz band used by TV broadcasters to be opened up for use by mobile operators at the beginning of 2017.

5.10 France

France planned to auction both digital dividend and 2.6 GHz band spectrum during May 2011, with paired spectrum being sold initially followed by unpaired spectrum. To underpin this, testing in both bandwidths was permitted during late 2010, and this concentrated on LTE FDD technology. For its part, Orange—which has now officially become the replacement name for France Télécom—conducted trials using FDD and TDD modes while Bouygues Télécom experimented with the 1,800 MHz band. In November 2010, the government announced that openness to virtual operators would be an important criterion when awarding 4G licences, adding in May 2011 that no operator would be permitted to win more than 15 MHz of digital dividend spectrum of which four 7.5 MHz lots were to be offered. This went down badly with the smaller operators which predicted that they would be shut out by Orange and Vivendi (SFR).

The auction process was launched in June with an application deadline of 15 September for 2.6 GHz spectrum and of 15 December for 800 MHz spectrum. Licences would be issued in 2011Q4 and 2012Q1 respectively. However, in June 2011, Bouygues petitioned the Council of State to strike down a condition in the

800 MHz licence that required its owner to compensate anyone whose enjoyment of digital terrestrial TV was spoilt by the mobile signal. For its part, Iliad (Free Mobile) went to court to postpone the auction on the grounds that it was prejudiced by the requirement that licences be paid for prior to their issue, but its action was unsuccessful.

In September, the four incumbents applied for 2.6 GHz licences via sealed bids combined with a decision as to whether or not to host MVNOs. Bouygues won 15 MHz paired at a cost of €228 million while Iliad won 20 MHz paired at a cost of €271 million and Orange won 20 MHz paired at a cost of €287 million, with all three agreeing to host MVNOs. For its part, SFR did not agree to host MVNOs and paid €150 million (the reserve price) for 15 MHz paired—it was speculated that it was reserving its firepower for the 800 MHz auction to follow. Hence, all of the available spectrum—unusually, there was no unpaired spectrum on offer—was allocated for €936 million ($1.28 billion) compared to a reserve price of €700 million (Cellular-news 2011a). The licences with specified bandwidth were handed over in October.

In mid-December, the same four operators applied for 30 MHz paired of 800 MHz spectrum which came with a coverage requirement amounting to 98 % of the population of mainland France within 12 years and 99.6 % within 15 years of the licence issue. The reserve prices for each block were set at a relatively high level which caused some adverse criticism to the effect that this played into the hands of deep-pocketed incumbents. In addition, the chosen method was a sealed-bid tender which ran the risk of bidders over-paying in order to be certain of getting a licence (the so-called 'winner's curse'). Bouygues won Block A (10 MHz paired) at a cost of €683 million while SFR won Blocks B and C (each of 5 MHz paired) at a cost of €1,070 million and Orange won Block D (10 MHz paired) at a cost of €891 million. The weakest operator, Iliad, was unsuccessful as was the norm elsewhere in Europe. However, once its 2.6 GHz network had achieved 25 % population coverage in January 2012, when it launched with HSPA+, it was able to roam onto SFR's 800 MHz network in rural areas not covered by its 2.6 GHz network (TeleGeography 2011d). Differences in the value of each block can partly be explained by the different amounts of interference to be expected given that it is the responsibility of operators to deal with interference issues. However, the conditions in the auction document do explain why SFR in particular ended up with the roaming agreement with Iliad insofar that it was SFR that won the blocks in the centre of the 800 MHz band. All licensees agreed to open their networks to 'full' MVNOs (those that terminate themselves), and also to commence their operations in rural areas (WirelessMoves 2012).

Orange, which launched a trial in Marseille in June 2012, pencilled in January 2013 for its launch of LTE using the 800 MHz and 2.6 GHz bands, as did Bouygues which launched a trial in Lyons and Villeurbanne in June 2012 using the 1,800 MHz band. For its part, Vivendi (SFR) initially set out to launch using the 800 MHz and 2.6 GHz bands, commencing in Lyon and Montpellier in January 2013. However, in September 2012, it stated its expectation that it would have the majority of the population covered with DC-HSPA+ by the year-end, providing a downlink of

42 Mbps, and hence it had decided to launch LTE at the end of November in Lyon (Telecom.paper 2012d). Shortly thereafter, Orange launched in Lille, Lyon, Marseille and Nantes albeit serving only corporate customers with a view to offering a consumer service in February 2013 and an extension of the network to ten cities in April (TeleGeography 2012x).

Meanwhile, in September 2012, a dispute over the location of towers and power limits for LTE signals in Paris was provisionally settled after months of controversy which had looked set to result in Paris being the last major city in France to access LTE. This meant that both Orange and SFR were able to launch in an initial district in January 2013.

At the year-end, Iliad took possession of 9.6 MHz of 900 MHz spectrum relinquished by the other incumbents, but assigned this for use on its 3G network. It admitted that it had made only a dozen base stations LTE-ready and was under offer from Vivendi. However, the antitrust authorities are adamantly opposed to such a takeover (TeleGeography 2013a).

In January 2013, it was alleged that Bouygues would not be allowed to re-farm its 1,800 MHz band spectrum for LTE until sometime in 2014 as its rivals were afraid that it would be able to launch on a nationwide basis before they were in a position to do so using their 800 MHz and/or 2.6 GHz licences—a comparison with the position in the UK is of interest here (TeleGeography 2013c). The matter was placed under review by the government which gave its permission in February subject to the payment of €60 million a year for 20 years—a very large sum in total compared to the norm for 2G licences in the band—and the fixing of a starting date for 4G services by the regulator (Telecom.paper 2013b). The regulator subsequently stated that the earliest date on which 1,800 MHz licences would be convertible to a technology-neutral basis would be May 2016, but if Bouygues or any other operator appealed this decision it would respond within 8 months. Also in January, the French antitrust authority ruled that licensees would be obliged to make wholesale offers to MVNOs prior to their own launches of LTE so that all parties could in principle launch simultaneously.

Further developments in January came respectively from Orange, which announced that it would be revamping its tariffs by April when it intended to launch in a further 15 cities—by which it meant that it would be raising the cost of LTE provision—and from SFR, which announced a partial launch in Paris.

The situation with respect to Bouygues became clearer after the regulator declared in March that, subject to ceding some 1,800 MHz spectrum—it would initially be obliged to reduce its holdings to a maximum of 23.8 MHz paired (as against its existing 26.6 MHz paired in urban areas), the same amount as Orange and SFR, with a subsequent stepped reduction to 20 MHz paired by end-May 2016—Bouygues would be allowed to launch in that band on 1 October. Orange and SFR were authorised to apply to re-farm the band if they so wished, although Iliad, owning no 1,800 MHz spectrum, would have to request an initial allocation in the band before proceeding. Although the regulator's intention is to end up with Orange, SFR and Bouygues each owning 20 MHz paired and Iliad owning 15 MHz paired of 1,800 MHz spectrum in 2016, the incumbents' response to a considerable increase in fees now that licences have become technology-neutral has yet to be determined. All told, the cost of a 20 MHz block of 1,800 MHz spectrum is due to rise to €64.6 million a year from €11.4 million a year, which is sufficiently high as to force operators to consider just how much spectrum they wish to utilise in this compared to other, cheaper, bands.

In practice, Bouygues set out to launch initially in eight cities in early May using the 2.6 GHz band (Sahota 2013). Hence, despite its favourable situation in respect of the 1,800 MHz band and in contrast with the position of EE in the UK (see Chap. 4), Bouygues had given its two main rivals roughly 6 months start in the provision of LTE services. Nevertheless, Orange is taking action to have Bouygues' right to use the 1,800 MHz band for LTE annulled, arguing that it would have behaved differently at the time of the 2.6 GHz and 800 MHz auctions if it had been made aware of the regulator's intentions concerning the 1,800 MHz band.

As for the 700 MHz band—in this case 694–790 MHz—the intention is to hold an auction in November 2015 with the intention of raising roughly $4 billion.

It may also be noted that the licensing process in French overseas departments (départements d'outre mer) is controlled by the French regulator Arcep. Arcep stated in June 2013 that it intended to sell licences direct to interested parties in La Réunion and Mayotte (islands off South-east Africa) and Guadeloupe, Martinique and French Guiana.

5.11 Germany

In early 2010, 41 blocks of spectrum no longer needed by broadcasters or the military in the 800 (790–862 digital dividend) MHz, 1,800 MHz, 2 GHz and 2.6 GHz bands, were offered for sale by auction. 60 MHz of 800 MHz spectrum was divided into six blocks of 5 MHz paired—for the background see Aetha (2011a). Vodafone and T-Mobile were allowed to bid for two blocks and KPN (E-Plus) and Telefónica (O₂) for three blocks, with new entrants allowed to bid for four blocks. But it was noted that, if the first two incumbents won two blocks each, little would be left over, so the other two incumbents took court action against allegedly unfair auction rules—however, they were unsuccessful. Only two new entrants expressed an interest, reflecting either a problem of financing or that the licences were all national rather than regional, and none bid. The auction started on 12 April and lasted for 224 rounds, concluding on 20 May at which point total bids amounted to €4.383 billion.

As shown in Table 5.1, of this sum €3.576 billion was bid for 800 MHz spectrum and the winners, all of whom won two blocks of 5 MHz paired, were Telefónica, Vodafone and T-Mobile—the latter two operators were capped at 10 MHz paired whereas Telefónica chose to limit its bid to that amount. In contrast, the weakest operator KPN came away empty-handed (although, in December, it was authorised to use its 900 MHz spectrum for LTE)—a result reflected elsewhere in Europe under similar circumstances (Aetha 2012). A further €359.5 million was bid for 58.8 MHz of 2 GHz. In addition, €344.3 million was bid for 2.6 GHz spectrum.

Table 5.1 Auction results. Germany. 2010

	800 MHz	1,800 MHz	2.0 GHz	2.6 GHz	€ million
Telefónica	10 paired	–	4.95 paired + 19.2 TDD	20 paired + 10 TDD	1,378
Vodafone	10 paired	–	4.95 paired	20 paired + 25 TDD	1,423
T-Mobile	10 paired	15 paired	–	20 paired + 5 TDD	1,298
KPN	–	10 paired	9.9 paired	10 paired + 10 TDD	284
Total MHz	60	50	39.6 paired + 19.2 FDD	140 paired + 50 FDD	
Total value €m	3,576	104	359	344	4,383

Overall, Vodafone paid €1,423 billion for 94.9 MHz of spectrum, Telefónica paid €1,379 million for 99.1 MHz, T-Mobile paid €1,299 million for 95 MHz and KPN paid €283.7 million for 69.8 MHz (Pyramid Research 2010).

In September 2010, having completed its trials prior to the auction, Vodafone launched in Helligendamm, the first of 1,500 communities that it intended its network to cover by March 2011. For its part, T-Mobile announced that whereas it intended to use its 2.6 GHz spectrum for rural coverage—which it, like the other winners in the band, was obliged to do with its new 800 MHz spectrum—it would deploy LTE using the 1,800 MHz band in high-usage areas. The 'Call & Surf Comfort via *Funk*' service was launched in 1,000 rural locations in December 2010 (Donegan 2010) and went live in Cologne using 1,800 MHz spectrum in June 2011, rapidly followed by a further 50 cities. Meanwhile, Telefónica launched trial networks using both 2.6 GHz spectrum in Munich and Halle and 800 MHz spectrum in Ebersberg and Teutschental. A commercial launch was pencilled in for July 2011 in rural areas using 800 MHz spectrum, but did not take place in urban areas until the July 2012 launch in Nuremberg and Dresden. The intention was to cover 100 cities by the year-end with a maximum downstream speed of 50 Mbps—barely faster than its HSPA+ service already available to 85 % of the population. However, at the end of August, the maximum LTE speed was doubled to 100 Mbps via a new 'LTE Speed service' which included the launch of an updated range of handsets (TeleGeography 2012q).

In a surprise move in August 2011, Vodafone announced that it intended to substitute LTE for its fixed-wire DSL services, with those customers wishing to retain DSL being sold on to other operators. This is expected to create significant financial benefits through the elimination of inter-connection payments for DSL and the use of a single LTE network for all data-heavy customers (Gabriel 2011a). For its part, KPN (E-Plus) announced a trial of TD-LTE in conjunction with China Mobile and ZTE in 2011Q1.

An interesting feature of the LTE roll-out was the licence requirement that rural areas lacking high-speed data services—so-called 'white spots'—should be the first to access LTE. Only once a given proportion of small towns had been provided for could the emphasis switch to larger towns and cities (Cellular-news 2012c). In May

2012, Vodafone, T-Mobile and Telefónica were deemed to have satisfied their rural coverage obligations in nine states with the full quota of 13 being achieved in November. For its part, KPN received permission to provide LTE using the 1,800 MHz band in June 2012, having previously been authorised to provide services to rural areas using its 900 MHz spectrum in December 2011 (TeleGeography 2012k). It completed tests using this band in Bonn in August.

In January 2013, Vodafone announced that it would have LTE coverage in 81 of the cities with at least 100,000 inhabitants by the end of March, bringing the total coverage to 50 million people.

In June, the regulator stated that it wanted to hold an auction for spectrum in the 900 MHz and 1,800 MHz bands during 2014 despite the fact that the existing licences in these bands are not due to expire until 2016. This directly contradicted the request by the incumbents that they be allowed to extend their licences until 2018 and be given permission to re-farm the spectrum for LTE on the grounds that they were strapped for cash as a result of ongoing investment in LTE licences and roll-outs (Total Telecom 2013).

5.12 Greece

An auction of 2G spectrum—largely consisting of licences that were due to expire in 2012—which could potentially be used for 4G as the auction was technology-neutral, was pencilled in for November 2011. However, only the three 2G incumbents expressed an intention to bid and each of these reserved a minimum amount of spectrum in the 900 MHz band, at a total cost of €182 million, in order to ensure the continuity of their existing services for the ensuing 15 years. In total, the auction offered 14 blocks in the 900 MHz band, each comprising 2.5 MHz paired with a cap of 15 MHz paired per operator, plus four blocks, each of 5 MHz paired, in the 1,800 MHz band (TeleGeography 2011c). The unreserved blocks raised a further €199 million, and overall OTE acquired four blocks of 900 MHz and two blocks of 1,800 MHz, Vodafone acquired six blocks of 900 MHz and two blocks of 1,800 MHz (at a cost of €169 million) and Wind acquired four 900 MHz blocks.

In August 2012, OTE became the first licensee to complete tests—it launched in Athens and Thessaloniki in November (Telecom.paper 2012j). Immediately there-after, Vodafone announced its intention to roll out the initial stages of its network before the end of 2012 and launched in Athens in December. However, this appears to have been a limited trial with a USB modem with the commercial launch taking place in June 2013 (TeleGeography 2013k). Subsequently, Vodafone announced a network-sharing agreement with Wind covering mainly rural areas. However, it was limited to 2G and 3G networks with Vodafone operating its LTE network independently throughout Greece.

An auction for 2.6 GHz band spectrum is next in line, to be followed during 2013 by an auction of digital dividend spectrum. Pending these developments, in February 2013, the regulator asked for comments on its plans to auction 140 MHz in the 3.4–3.8 GHz band. This could potentially be used for LTE-Advanced or WiMAX

Table 5.2 Spectrum holdings of incumbents. Hungary. 2011

Band	Telenor	T-Mobile	Vodafone	Available
900 MHz	8 MHz paired	8 MHz paired	8 MHz paired	10.8 MHz paired
1,800 MHz	28.8 MHz paired	15 MHz paired	15 MHz paired	15 MHz paired
2.1 GHz	15 MHz paired	15 MHz paired	15 MHz paired	15 MHz paired

802.16e. The 3,410–3,500 MHz and 3,510–3,600 MHz bands have been licensed to three parties—Craig Wireless, Wind Hellas and OTE—since 2000 and there is not expected to be any interest in spectrum beyond 3,600 MHz.

5.13 Hungary

Vodafone was the first incumbent to announce plans to launch LTE, pencilling in 2013, but Telenor and T-Mobile (Magyar Telekom) wanted to launch earlier once spectrum was made available. T-Mobile went so far as to pencil in 1 January 2012 for a launch on the back of a testing phase beginning in October 2011, and this duly took place in Budapest using existing spectrum in the 1,800 MHz band. According to the GSA (Global mobile Suppliers Association 2012), Telenor launched in the same band in July 2012 using the brand 'Hipernet 84', with coverage in 60 towns. However, the relevant Telenor press release (Telenor 2012), gave the impression that the maximum downlink of 84 Mbps was being provided by a variant of HSPA+ rather than LTE.

In July 2011, the government authorised an auction of unused GSM frequencies for possible LTE or WiMAX usage. The winner of the largest (5 MHz paired) block in the 900 MHz band one-round tender—initially expected to be restricted to new entrants provided they were prepared to pay a minimum fee – would be entitled to bid for three additional 5 MHz paired blocks in both the 1,800 MHz and 2,100 MHz bands, while other 1 MHz or 0.8 MHz blocks (either paired or unpaired) would be available to all-comers subject to an individual operator cap of 7.8 MHz of the total of 10.8 MHz on offer. The applicants were Telenor Hungary, Vodafone, T-Mobile, a consortium comprised of Magyar Posta, MVN and MFB trading as MPVI, RCS&RDS (Romania) and the Viettel Group (Vietnam), but the latter two were disqualified from bidding (TeleGeography 2012b).

It is worth noting the spectrum holdings of the incumbents at this point in time in Table 5.2 since, for example, they held enough 900 MHz spectrum for voice-only purposes but would need additional bandwidth if the band was to be re-farmed for high-speed broadband.

The 900 MHz band spectrum, auctioned in January 2012, was awarded to the remaining potential new entrant, the state-owned MPVI consortium (the 5 MHz 'A' block covering a spectrum band suitable for voice and data at a cost of $44.9 million) and to incumbents T-Mobile (2 MHz at a cost of $46.4 million), Vodafone (2 MHz at a cost of $66.6 million) and Telenor (1.8 MHz at a cost of

$31 million)—the latter allocations constituting E-GSM spectrum. The 'B' and 'C' block allocations to the incumbents are insufficient by themselves to roll out a LTE network and are suitable only for rural coverage, but it does mean that the incumbents have give or take the minimum 10 MHz paired in the band to sustain re-farming for broadband while awaiting the issue of licences in the 800 MHz band. However, the 'A' block decision was appealed on the grounds of unfair advantage, if only to delay the roll-out of the new operator, and the licence was annulled by a court in September (Cellular-news 2012e). As the court denied the right to appeal, it looked very much as though MPVI was dead in the water. Slightly oddly, the case nevertheless ended up in the Supreme Court which confirmed the lower court's decision (Cellular-news 2013a). MPVI suspended its operations in April in order to minimise costs.

In February 2013, ZTE announced the launch of Telenor's LTE network in the 1,800 MHz band, seemingly confirming that all previous developments had related to HSPA+. The new network covered 76 towns (4G-Portal 2013).

5.14   Iceland

In February 2013, the regulator announced that it would be auctioning 60 MHz in the 800 MHz band, comprising 10 MHz paired licensed for 25 years (Licence A) at a reserve price of $786,000 plus four 10-year licences (Licences B, C, D and E) each of 5 MHz paired at a reserve price of $157,000, together with 50 MHz in the 1,800 MHz band comprising five 10-year licences (Licences F, G, H, I and J) each of 5 MHz paired at a reserve price of $39,300. The reserve prices were equivalent to $0.05/MHz/pop. The auction would take place the following month. Whereas the 1,800 MHz licences would be technology-neutral, the 800 MHz licences would have data transfer speed and coverage obligations attached. Provided a 3G licence was already held—as in the case of Novator, Síminn and Fjarskipti (Vodafone Iceland)—the winner of licence A would be obliged to provide a 2 Mbps downlink to 99.5 % of homes and businesses by end-2016 and a 30 Mbps downlink with identical coverage by end-2020. For other licensees, the coverage requirement would be reduced to 93.5 % by various deadlines.

3G licensees would be capped at 20 MHz paired overall—unless one was awarded Licence A in which case it would be allowed to win a further 5 MHz paired in the 1,800 MHz band—and other bidders at 20 MHz paired in the 800 MHz band plus 15 MHz paired in the 1,800 MHz band. Technically, this meant that an incumbent plus a new entrant could walk away with the whole of the spectrum (Jóhannsson 2013).

The licences were won by Novator, Síminn and Fjarskipti together with new entrant 365 Media which offered the highest bids totalling $1 million for bands C and J. The other bids were worth only $875,000 in total so it is evident that 365 Media was determined to use this opportunity to break into the mobile market. Novator launched in the Reykjavik area in April 2013 (Telecom.paper 2013d).

5.15   Ireland

The regulator announced in March 2012 that it would shortly be conducting the much delayed auction of 28 blocks (140 MHz paired) of spectrum with licences covering the period February 2013 to February 2030, albeit divided into two stages with the first ending in July 2015 due to the expiry of GSM licences. This auction would encompass 900 MHz (880–915 MHz paired with 925–960 MHz) and 1,800 MHz (1,710–1,785 MHz paired with 1,805–1,880 MHz) licences which could then be re-farmed to provide 3G and LTE (TeleGeography 2012e), as well as the digital dividend spectrum (791–821 MHz paired with 832–862 MHz) which would become available in October 2012. The auction would more than double the existing spectrum assigned in these bands. Spectrum caps would apply to both individual and joint-entity bidders. The reserve prices, payable partly up-front and partly by way of annual fees, were as follows: €20 million per 5 MHz paired in the 800 MHz and 900 MHz bands and €10 million per 5 MHz paired in the 1,800 MHz band (TeleGeography 2012f).

In June, the requirement that successful bidders cover at least 70 % of the population within 3 years was criticised on the grounds that this would be achievable with a network covering only 10 % of the landmass. The regulator responded that there was no reason for operators to restrict coverage to the largest urban areas.

In July 2012, Vodafone and Hutchison (H3G Ireland) announced that they had signed an agreement to set up a 50/50 joint venture to manage their cell sites and transmission networks, in the process decommissioning any duplicate facilities. However, they would continue to manage independent core networks and spectrum.

The results of the auction in mid-November were as shown in Table 5.3 (Cellular-news 2012h). All four successful bidders were 2G and/or 3G incumbents. All of the available 140 MHz paired was sold for the period 2015–2030 together with 95 MHz paired for the period 2013–2015. The total amount raised was €854.5 million ($1,091 million).

Also in November, Vodafone announced that it would be launching in June 2013 or thereabouts. This was slightly at odds with the subsequent claim by Samsung that it had been contracted by Hutchison to build the first LTE network in Ireland with a proposed launch by H3G in Dublin in August 2013 (Cellular-news 2013c). It is possible that around that time, or soon thereafter, H3G will have become a new entity having agreed the takeover of Telefónica subject to regulatory clearance. If cleared, the enlarged entity will have almost the same market share (roughly 40 %) as Vodafone, and Hutchison has stated that heavy investment in 4G will form part of the strategy designed to displace Vodafone as market leader.

Meanwhile, Imagine Communications claims to have launched a 4G WiMAX network.

Table 5.3 Auction results. Ireland. 2012

Band	Licence	Hutchison	Eircom	Telefónica	Vodafone
(MHz)	(Period)	('3')	(Meteor)	(O₂)	
800	2013–2015	–	20 MHz	20 MHz	20 MHz
900	2013–2015	10 MHz	10 MHz	20 MHz	20 MHz
1,800	2013–2015	20 MHz	20 MHz	–	30 MHz
800	2015–2030	–	20 MHz	20 MHz	20 MHz
900	2015–2030	10 MHz	20 MHz	20 MHz	20 MHz
1,800	2015–2030	40 MHz	30 MHz	30 MHz	50 MHz
Upfront fee €m	–	51.1	144.8	124.9	160.9
Spectrum fee €m	–	53.9	99.6	99.6	119.6

5.16 Italy

Italy had already authorised re-farming of the 900 MHz and 1,800 MHz bands and this was to be followed by the auctioning of additional spectrum in the 1,800 MHz and 3G bands. However, plans to auction digital dividend spectrum in the 800 MHz band in September 2011 were initially cast into doubt when the spectrum owners refused to accept a payment of 10 % of the auction proceeds in return for releasing the spectrum. The 2.6 GHz band was also pencilled in to be offered via an auction open to all-comers. Vodafone commenced trials in 2009 and Telecom Italia in 2010.

In July 2011, five applications were lodged for an auction of spectrum in all of the above bands—in total 240 MHz was on offer in the 800 MHz, 1,800 MHz and 2.6 GHz bands plus one block of 15 MHz TDD in the 3G band. The 15-year, technology-neutral licences were subject to spectrum caps for individual licensees of 25 MHz paired in the 800 MHz/900 MHz bands, 25 MHz paired in the 1,800 MHz band and 55 MHz in total (FDD + TDD) in the 2.6 GHz band. The 800 MHz blocks, other than the lowest which had a usage restriction attached, came with a coverage requirement that a 2 Mbps downlink had to be provided to all towns with fewer than 3,000 inhabitants. Linkem (which intends to switch from WiMAX to LTE using the 3.5 GHz band) expressed an interest in addition to 2G licensees Telecom Italia, Vodafone, Hutchison and Wind. The bids from authorised parties had to be lodged by the end of August, and in the event only the mobile operators applied. The government set the reserve prices at a total of €655 million, hoped to raise €2.4 billion in total and managed to sell the 800 MHz licences by themselves for €3 billion. The six blocks were won by Telecom Italia and Vodafone—each of which also won one 1,800 MHz block and three 2.6 GHz blocks at a total cost of €1.26 billion—and Wind which also won four 2.6 GHz blocks at a cost of €1.12 billion. Hutchison ('3' Italia) expressed fears that it would struggle to roll out its 4G network, and in the event it did not win a 800 MHz licence, instead buying one 1,800 MHz block and four 2.6 GHz blocks at a cost of €305 million as shown in Table 5.4.

Table 5.4 Auction results. Italy. 2011

	800 MHz	1,800 MHz	2.6 GHz	€ million
Telecom Italia	10 paired	5 paired	15 paired	1,260
Vodafone	10 paired	5 paired	15 paired	1,260
Wind	10 paired	–	20 paired	1,120
Hutchison	–	5 paired	10 paired	231
			30 unpaired	
Total MHz	60	30	120 paired + 30 TDD	74
Total value €m	2,962	477	432 paired + 74 TDD	3,945

Despite sharp criticism to the effect that it had not been the best-managed of auctions, the government claimed that the amount raised (€3.94 billion) was 'extraordinary'. Given the poor economic climate at the time and the antagonism of the regional TV operators that had been forced to vacate the spectrum for what they considered to be inadequate compensation, this seemed a reasonable conclusion to draw. Partly as a consequence of the relatively high bids, the initial effect for Wind was to cause a downgrade in its debt rating on the grounds that it had become more susceptible to adverse economic conditions. The 800 MHz blocks were reclaimed at the year-end and handed to the three successful bidders in January 2013.

Hutchison intended to be the first to launch, using re-farmed 1,800 MHz spectrum, before the end of 2012, with total spending on its roll-out amounting to €1 billion over a 3-year period. The intention was to achieve nationwide DC-HSPA+ coverage alongside pockets of LTE (Gabriel 2012a). It would be followed by Wind which set out to launch LTE in all major cities by 2013. Meanwhile, Vodafone also committed itself to a major launch in 2013 after spending €1 billion on its roll-out. However, in practice, it was Telecom Italia which was reported to have achieved the first launch in October 2012 in Rome, Milan, Turin and Naples (Telecom.paper 2012f), followed later in the month by Vodafone which launched in Rome and Milan with a further 18 cities to follow by March 2013 (Telecom.paper 2012h). Hutchison launched days later, albeit only in the village of Acuto, and it was then claimed that Telecom Italia had in fact launched a further few days after Hutchison (TeleGeography 2012w). Either way, the initial three launches were close together and none of the networks were expected to achieve significant coverage until 2013.

In April 2012, the government announced that it would no longer be allocating digital dividend spectrum on the basis of a beauty contest but would instead conduct an auction, with interested parties to be registered with the regulator within 4 months.

Also in April, highly-indebted Telecom Italia revealed that Hutchison Whampoa was interested in becoming its largest shareholder by folding into it 3 Italia and buying shares from existing shareholders (Reuters 2013). However, there were at least two potential roadblocks to progress with such a deal, namely the need to hive off Telecom Italia's fixed-wire network and the opposition of Telefónica which held a large stake in Telecom Italia's controlling shareholder Telco, so it came as no great surprise when talks were suspended in July.

5.17   Jersey

In August 2009, the JCRA recommended to the UK's Ofcom, the ultimate regulator, that a little-known entity, Clear Mobitel, should be awarded a test licence for LTE in the 2.6 GHz band. It then withdrew its recommendation in March 2011, but Clear Mobitel was successful in having that decision struck down in the Royal Court of Jersey in September (ClearMobitel 2011).

5.18   Latvia

TeliaSonera launched using its 2G spectrum in May 2011. In January 2012, five blocks of spectrum in the 2.6 GHz band were offered by auction at a minimum price of $372,000 per 20 MHz paired block. However, only four applications were made, by TeliaSonera (which paid $1.44 million for 2,500–2,520 MHz paired with 2,620–2,640 MHz), Tele2 (which paid $1.38 million for 2,520–2,540 MHz paired with 2,640–2,660 MHz), Bité (which acquired 2,540–2,560 MHz paired with 2,660–2,680 MHz), and cable operator Baltkom (which paid $0.25 million for 2,560–2,570 MHz paired with 2,680–2,690 MHz). The 2,570–2,620 MHz licence was not sold. The licences are operational from 1 January 2014 and run for 15 years (TeleGeography 2012a).

In October 2012, the regulator awarded spectrum in the 2,300–2,330 MHz band to TeliaSonera and in the 2,330–2,360 MHz band to Bité at a cost per licence of $205,000.

5.19   Lithuania

Although TeliaSonera launched in April 2011 using 2G spectrum, the auction for 2.6 GHz spectrum did not take place until March 2012. The regulator had announced that there would be three 15-year licences, each comprising 20 MHz paired in the 2,500–2,560 MHz and 2,620–2,680 MHz bands, with licensees required to cover at least 50 % of five cities within 3 years and at least 50 % of 15 cities within 5 years (Cellular-news 2011c). The three incumbents, TeliaSonera, Bité GSM and Tele2, were successful—no fees were published—and Tele2 announced that it stood ready to launch within 1 month of getting its licence. The launch duly took place in five cities in March 2013 (TeleGeography 2013f).

In June 2013, the regulator stated that there would be a single auction in August for 800 MHz spectrum with success to be based solely upon the size of the bids. One lot of 10 MHz and four lots of 5 MHz would be made available and one winner would be obliged to cover 95 % of rural households by 2020. A further auction for spectrum in the 2.6 GHz band will be held at a later date.

5.20 Luxembourg

Orange completed a 6-month trial during 2010. However, the trial demonstrated that there are likely to be serious issues of interference with networks in neighbouring countries if the 2.6 GHz band is to be used (TeleGeography 2012j).

When Belgacom's 15-year 2G licence was renewed in June 2012, it was extended to include permission to launch LTE. Orange and state-owned EPT (LuxGSM) received their 4G licences at the same time, together with the additional 1,800 MHz spectrum awarded to all three incumbents. Belgacom (trading as Tango) launched commercially with 60 % population coverage in October 2012, intending to increase this to 90 % by the year-end (Globaltelecomsbusiness 2012). This triggered a response from Orange to the effect that it would also launch by the end of the month providing up to 50 Mbps downstream and 20 Mbps upstream across 70 % of the country, rising to 90 % during 2013 (TeleGeography 2012t).

5.21 Macedonia

In November 2011, the regulator invited international operators to participate in a tender for digital dividend spectrum which would proceed provided at least three intended to bid. The minimum fee would be $3.3 million and the winner(s) would be required to build 80 base stations. However, no interest was forthcoming in October 2012. The date for awarding the licence was subsequently set back to 1 September 2013.

In May, the 1,710–1,880 MHz band was added to the above and the term of the licences fixed at 20 years from 1 September 2013. Applicants would need to be supplying at least 300,000 customers at the time that they tendered and have declared total assets from telecoms activities of at least €10 million ($13 million) in both 2011 and 2012. The winners of the beauty contest would be selected primarily on the basis of the amount bid although account would also be taken of their detailed roll-out projections (Telecom.paper 2013f). Population coverage targets would apply from the issue date of the licences as follows: 20 % within 2 years; 40 % within 4 years; and 70 % within 6 years.

Each of the three complete licences consisted of 10 MHz paired within 791–821 MHz paired with 832–862 MHz together with 15 MHz paired within 1,740–1,785 MHz paired with 1,835–1,880 MHz, priced at €10.3 million, but applicants could also opt to bid for sub-blocks, each of 5 MHz paired, of which there were accordingly five per complete licence. In this case, the price per sub-block was set at €2 million. A complete licence could only be won if there were at least two bidders for the licence and any applicant could bid for two (but not three) complete licences (TeleGeography 2013j). In June it was alleged that three bidders had come forward, namely incumbents T-Mobile (in which T-Mobile owns a minority stake via Magyar Telekom), Telekom Slovenije (ONE) and Telekom Austria (Vip Operator).

In July 2013, it was announced that Telekom Austria had acquired a complete licence for €10.3 million ($13.1 million) and was obliged to launch by 1 August 2014 (TeleGeography 2013o). It was confirmed shortly thereafter that the two other bidders had bought similar licences on the same day for the same fee.

5.22 Monaco

Monaco Telecom was licensed in November 2011. It contracted with Huawei with a view to a launch in September 2013 to be followed by a roaming agreement with Orange in France.

5.23 Montenegro

Telenor commenced a trial in Podgorica in November 2011 and launched in four cities in November 2012 (SeeNews 2012).

Velatel is rolling out a TD-LTE network using 40 MHz in the 3.5 GHz band.

5.24 Netherlands

An initial trial auction of 190 MHz of spectrum in the 2.6 GHz band was held in December 2009, with the auction proper scheduled for April 2010. Unusually, the regulator reserved certain blocks, totalling 40 MHz paired and 55 MHz unpaired, for new entrants whereas incumbents KPN, T-Mobile and Vodafone were each subject to spectrum caps. Of necessity, these restrictions meant that reserve prices were set at a relatively modest level. In the event, and most unusually for Europe, the unpaired spectrum was ignored and only €2.6 million was raised. New entrants Tele2 and Ziggo4 (a joint venture between cable operators Ziggo and UPC) each won 20 MHz paired, while KPN and Vodafone acquired 10 MHz paired and T-Mobile acquired 5 MHz paired. The licences are valid until 2030 but an initial launch was required within 2 years of licence receipt. In practice, in order to meet the launch date deadline, KPN, Tele2 and Ziggo4 all launched within a matter of days at the beginning of May 2012 (Telecom.paper 2012a), with Vodafone and T-Mobile following suit shortly thereafter. Not surprisingly, all of the operators made do with providing services to a limited group of business customers in no more than a handful of cities and there is some disagreement as to whether this merits being treated as the launch date for LTE.

At the end of October 2012, a further (postponed) auction was pencilled in which would include the existing 2G holdings of incumbents which were due to expire in February 2013 and were not extendable—at least in theory, given that the

Table 5.5 Auction results. Netherlands. 2012

Lot	Licences	Band	MHz	Reserve €m	Winners (MHz)
A1	2	800	20	35	Tele2
B	4	800	40	35	KPN (20), Vodafone (20)
C	7	900	70	28.9	KPN (20), T-Mobile (30), Vodafone (20)
D	14	1,800	140	125	KPN (40), T-Mobile (60), Vodafone (40)
E	2	2,100	20	0.8	KPN (10), Vodafone (10)
F	2	1,900	14.8 TDD	0.6	T-Mobile
G	10	2,600	50 TDD	0.6	KPN (25), T-Mobile (25)
H	1	2,600	5 TDD	0	KPN

government allowed incumbents to apply for an extension in September 2012—together with 60 MHz of paired spectrum from the digital dividend band and the remaining spectrum in the 2.1 GHz and 2.6 GHz bands. 10 MHz paired in the 800 MHz band would be reserved for a new entrant—Tele2 was quick to register its interest—as well as 5 MHz paired in the 900 MHz band (TeleGeography 2012r). As in the 2.6 GHz auction in 2010, a coverage requirement of 20 km^2 after 2 years and 200 km^2 after 5 years would apply with licences lasting until 2030. The precise offer, with each licence (bar F) comprising 5 MHz paired or 5 MHz unpaired (Lots G and H), was as shown in Table 5.5.

Lot C comprised the entire allocation of 900 MHz bandwidth as no potential new entrants applied for the ring-fenced quota. Lot E comprised the 3G licence returned by Telfort after being taken over by KPN with an expiry date of 2017 (to match other 3G licences). Lot H was a free-of-charge bonus for one winner of the Lot G licences (TeleGeography 2012v). Lots F, G and H would be licensed from 1 January 2013 for 17 years.

Bidders were expected to comprise incumbents KPN, Vodafone and T-Mobile together with Tele2 (an existing MVNO using the T-Mobile network) and Ziggo/UPC (bidding as Zum). In the event, Zum proved incapable of bidding competitively during the entire 7 weeks auction so Tele2 became the only new entrant, albeit only in the sense that it was now in a position to roll out a network of its own (for the provision of LTE). Tele2 paid €160.8 million for its licences compared to the €911 million paid by T-Mobile, the €1.35 billion paid by KPN and the €1.38 billion paid by Vodafone. When account is taken of the fact that the government expected to raise only €480 million in total, the above amounts appear to be irrationally high—so high, indeed, that KPN subsequently announced that it would be cancelling its dividend (Gabriel 2012b). It also stated that it would use the 800 MHz spectrum to launch in the north of the Randstad in February 2013 with nationwide coverage available at some point in the summer.

It is of interest to reflect upon the competitive landscape in the Netherlands after the above auctions. As might be expected, the three incumbents remain dominant. Across the five main bands (800/900/1,800/2,100/2,600 MHz) T-Mobile now has the most paired spectrum (around 270 MHz) with much the largest holding in the

1,800 MHz band (124 MHz), while Vodafone has the smallest holding (just over 200 MHz). In contrast, Tele2 and Ziggo/UPC each have less than 100 MHz (Telecom.paper 2012k). Which spectrum bands will be used for LTE, and how much investment will take place given the costs of the licences remains to be seen.

For its part, KPN announced that it would be launching in the Amsterdam region in February 2013, using the 800 MHz band, and going nationwide by mid-2014. The maximum downlink would be 50 Mbps and the expected average around 20 Mbps (Telecom.paper 2013a). It appeared to be treating this as the formal commercial launch. Vodafone also launched in Randstad in February. In contrast, T-Mobile claimed that it would be launching in 2013Q4 using the 1,800 MHz band in and around Amsterdam while Tele2 stated that it would be delaying its launch for several quarters while it concentrated upon improving the quality of the network.

5.25 Norway

The results of the 2.6 GHz band auction—the first in Europe—were announced in Norway in November 2007. This was a complicated auction as it included both paired and unpaired spectrum in the 2.6 GHz band, divided into regional blocks, as well as unpaired spectrum in the 2,010–2,025 MHz band—for a full band plan and results see Global View Partners (2009: Tables 14 and 15) and Aetha (2012: p. 33). Given the existence of only two mobile incumbents, TeliaSonera and Telenor, each was able to acquire a practical 20 MHz paired in the 2,500–2,540 MHz paired with 2,620–2,660 MHz bands (with the exception of the 15 MHz paired acquired by Telenor in Oslo). Telenor also won 40 MHz of the 60 MHz available (2,540–2,570 MHz and 2,660–2,690 MHz) for use in either a FDD or TDD format. Craig Wireless won 50 MHz of unpaired spectrum (2,570–2,620 MHz) on a nationwide basis (comprising 5 blocks in 6 regions), while Inquam won the 2,010–2,020 MHz band licence in all six regions. Given that other WiMAX operators Hafslund Telekom and Arctic Wireless also won unpaired spectrum, with Hafslund mopping up the few unsold regional licences in 2008, this indicated a willingness to exceed the standard unpaired spectrum allocation agreed upon by European Union regulatory bodies of 50 MHz unpaired.

The unusually low total of €29 million was raised, with paired spectrum actually selling for less than unpaired spectrum in terms of dollars per MHz per pop other than in Oslo, but this presumably reflected in part the unusual market structure. Interestingly, when Craig Wireless sold-on half of its spectrum in February 2009, it received less than half of what it had paid (Global View Partners 2009). TeliaSonera launched in Oslo in December 2009 but had accumulated a meagre 5,700 LTE subscribers by mid-2012. For its part, Telenor did not launch until October 2012, initially in 11 cities with a view to 90 % population coverage by 2015 (Hibberd 2012). It introduced handsets in December.

A further auction in November was next on the agenda, comprising 20-year technology-neutral licences in the 1,920–1,980 MHz band paired with the 2,110–2,170 MHz band. Two blocks of 4.8 MHz paired plus seven blocks of

5 MHz paired were on offer with bidders restricted to a maximum of 20 MHz paired. TeliaSonera and Telenor each won four blocks while Tele2 (Mobile Norway) won one block to add to the three that it already controlled. The other two (unidentified) bidders withdrew before recording a bid (Cellular-news 2012i) with the result that the auction only lasted for one round and each block sold at the floor price of $866,482. A meagre total of $7.8 million was raised.

An auction of spectrum in the 800 MHz band is well advanced although a specific date has yet to be fixed. A 10 MHz paired block is to be offered in the centre of the band with the winner obliged to provide coverage to 98 % of the population and a minimum downlink of 2 Mbps within 5 years of licence receipt. Four further blocks of 5 MHz paired will be on offer but the obligation attached will merely be 40 % population coverage within 4 years. Either some or all operators will be restricted to a cap of 10 MHz paired. 15 MHz paired in the 900 MHz subject to an individual operator cap of 15 MHz paired together with 55 MHz paired in the 1,800 MHz band subject to an individual operator cap of 20 MHz paired will also become available.

5.26 Poland

In late 2009, the incumbent operators boycotted a tender for a single 2.6 GHz licence providing 50 MHz of unpaired spectrum on the grounds that the licence terms were too onerous. It was won in November by Aero2 (Mobyland) with Milmex Systemy Komputerowe the only other bidder. Aero2 proceeded to build a FDD network in conjunction with CenterNet—these have now merged into a single company—but chose to use its existing 1,800 MHz spectrum. In March 2011, Polish satellite DTH provider Cyfrowy Polsat initiated a trial of LTE using the Aero2/CenterNet network followed by a commercial launch in August (acting as a MVNO). Aero2 set out to launch the world's first combined FDD/TDD network built by Huawei (TeleGeography 2011b), and launched the TD-LTE network using its 2.6 GHz spectrum in May 2011 thereby enabling the combined network to also be launched.

In 2010, the regulator announced a tender for two 35 MHz blocks in the 2.6 GHz band. This was not well received given the small number of licences and the large amount of spectrum per licence. Pending the auction, Orange and Netia P4 proposed a joint LTE network (which was approved by the regulator) while Polkomtel began testing in September 2010 and T-Mobile built a LTE-ready network. In early December 2011, Polkomtel—which is linked to both Aero2 and Cyfrowy Polsat through the holdings of Zygmunt Solorz Zak—announced the launch of its service, initially covering 22 % of the population, using the Aero2 network—in this case combining its HSPA+ 900 MHz and LTE 1,800 MHz spectrum—as a MVNO. In November, it launched a dongle/router-based service with a theoretical maximum downlink of 150 Mbps—for full details see Telecom.paper (2012g).

Meanwhile, in August 2012, the regulator announced its intention to advance with all speed towards an auction of spectrum in the 1,800 MHz band with further auctions in the 800 MHz band (791–821 MHz paired with 832–862 MHz) and 140 MHz in the 2.6 GHz band to follow in 2013. It went on to specify that there would be five 15-year licences on offer (expiring on 31 December 2027) in the 1,730–1,755 MHz band paired with the 1,825–1,850 MHz band, with an applicant allowed to bid for up to three licences. A commercial launch would be required within 1 year of receipt of a licence. The main concern would be to ensure competition, taking into account the amounts offered and the commitment to invest in new infrastructure (TeleGeography 2012p). Six applications were forthcoming in October from incumbents Orange, Polkomtel, T-Mobile (PTC) and Netia P4 together with fixed-wire operator Sferia (a partner of Aero2) and transmission wholesaler EmiTel. All bar Polkomtel and Sferia, which expressed interest in two licences, bid for the maximum three licences. The auction concluded in mid-February. Netia P4 was awarded three licences for €119 million and T-Mobile the other two for €109 million. The total raised exceeded initial expectations by over 50 % (Adamowski 2013). However, the licences were not handed over until June.

Progress in respect of the rest of the 800 MHz band has been stymied by the reluctance of the army to hand over its spectrum in the 790–806 MHz, 837–838 MHz and 846–847 MHz bands for civilian use, which they can legally delay until 2017. A further difficulty stems from the fact that Polkomtel, EmiTel and Sferia filed complaints about the outcome of the 1,800 MHz auction, claiming that the technical grounds cited as the reasons for their respective failures to win spectrum were misguided and possibly illegal. Meanwhile, Netia P4 and T-Mobile are trying to persuade the regulator to postpone any further auction on the grounds that their investments in roll-outs will leave them with too little cash to bid competitively, but they are unlikely to succeed.

In August 2011, broadband provider Milmex partnered with Airspan to launch LTE using existing 3.5 GHz spectrum originally utilised for WiMAX. It hopes to progress to LTE-Advanced after 2015 (TeleGeography 2012h). Sferia also intends to roll out a LTE network using the 850 MHz band on its own account after years of delay caused by legal and regulatory complications.

5.27 Portugal

The plan was to hold an auction in June 2011 involving spectrum in the 450 MHz, 800 MHz, 900 MHz, 1,800 MHz, 2.1 GHz and 2.6 GHz bands, all of which were considered to be suitable for LTE. Each 800 MHz block came with a requirement to cover up to 80 parishes without an existing broadband coverage within a specified timeframe (although existing 900 MHz spectrum could also be utilised to this end). In anticipation, Portugal Telecom (TMN) conducted a series of trials commencing in mid-2010. The auction was postponed in order to rewrite the auction rules, and in October 2011 it was announced that applications had been received from Portugal Telecom, Sonae.com (Optimus), Vodafone, Zapp and the CPMCS. The auction commenced in late November and the winners were as set out in Table 5.6. Both

Table 5.6 Auction results. Portugal. 2011

	800 MHz	900 MHz	1,800 MHz	2.6 GHz	€ million
Portugal Telecom	10 paired	–	14 paired	20 paired	113
Vodafone	10 paired	5 paired	14 paired	20 paired + 25 TDD	146
Sonae.com	10 paired	–	14 paired	20 paired	113
Total MHz	60	10	84	120 FDD + 25 TDD	

Portugal Telecom and Sonae.com paid the minimum cost of €113 million for nine blocks in three bands. The total spectrum available was sold only in the 800 MHz band and there was no new entry—for details see http://www.anacom.pt. One factor accounting for this was the existence of an overall spectrum cap imposed across all of the available bands, which meant that operators ignored the bandwidth with the lowest value relative to its price—in this case the unpaired 2.6 GHz spectrum.

Vodafone pledged to launch in 13 cities using the 2.6 GHz band as soon as it received its licence, while Portugal Telecom, which launched simultaneously in March 2012 with 20 % population coverage, was allegedly able to raise that figure to 80 % by mid-May (TeleGeography 2012i) after it gained access to digital dividend spectrum. However, Vodafone has questioned whether this figure is realistic and Portugal Telecom was told to cease making any claims to the effect that its network had the widest coverage in Portugal until the matter had been officially investigated. For its part, Sonae.com claimed that it would launch the very same day although in practice there was a slight delay.

In May 2013, Sonae.com and Portugal Telecom stated that they had achieved a downlink of 300 Mbps and an uplink of 70 Mbps using LTE-Advanced equipment supplied by Huawei Technologies. However, no devices were as yet available to take advantage of these speeds.

5.28 Romania

In January 2012, a public consultation was opened over the use of spectrum in the 830–862 MHz, 1,747.5–1,785 MHz, 1,842.5–1,880 MHz and 2,500–2,690 MHz bands, all in partial use at the time by the Ministry of Defence. However, if these bands were to be auctioned, they had first to be re-farmed at a total cost of roughly €145 million.

In March, the regulator announced an auction planned for September 2012 which consisted of 42 blocks of paired spectrum, each comprising 5 MHz paired: six blocks in the 800 MHz band at a minimum of €35 million each, seven blocks in the 900 MHz band at a minimum of €40 million each (both bands being sold with population coverage conditions attached), 15 blocks in the 1,800 MHz band at a minimum of €10 million each and 14 blocks in the 2.6 GHz band at a minimum of €4 million each. In addition, three unpaired blocks of 15 MHz were to be offered in the 2.6 GHz band at a minimum of €3 million each. At these low prices, interest

Table 5.7 Auction results. Romania. 2012

	800 MHz	900 MHz	1,800 MHz	2.6 GHz	€ million
OTE	5 paired	10 paired	25 paired	10 paired	180
Vodafone	10 paired	10 paired 12.5 paired S/T	30 paired 15 paired S/T	15 TDD	229
Orange	10 paired	10 paired 12.5 paired S/T	20 paired 15 paired S/T	20 paired	227
RCS&RDS	–	5 paired	–	–	40
2K Telecom	–	–	–	30 TDD	7
Total	50	70	150	105	
Total S/T	–	50	60		

was expected from international telcos, with AT&T, China Mobile and Telenor in the frame but in the event none applied.

The 15-year licences—the first to be auctioned in Romania—would commence on 15 April 2014. To facilitate this, the blocks in the 900 MHz and 1,800 MHz bands currently used by Orange and Vodafone—10 blocks of 2.5 MHz paired in the 900 MHz band and six blocks of 5 MHz paired in the 1,800 MHz band—would be offered covering the period from January 2013 until April 2014 (Cellular-news 2012b).

Five applications were forthcoming in August for the auction in September. The applicants comprised incumbents Vodafone, OTE, Orange and RCS&RDS plus new entrant 2K Telecom (majority owned by Alexandru Ghita). Because of the nature of the initial interest it was decided to hold a preliminary round of bidding with the minimum bid set at the reserve price and with each applicant specifying the number of blocks it wanted to buy in each spectrum band at that price. If demand exceeded supply in any band then a further round of bidding would take place until there was no longer any excess demand. The auction results were as shown in Table 5.7—S/T refers to the period 1 January 2013 to 5 April 2014.

OTE, RCS&RDS and 2K Telecom agreed to host MVNOs in return for reduced coverage obligations—Orange and Vodafone were obliged to prioritise 676 under-served rural communities. It is worth noting that one 5 MHz block in the 800 MHz band and eight 5 MHz blocks in the 2.6 GHz band were left unsold, indicating once again that the latter band tends not to be popular if spectrum is available in lower bands. In terms of spectrum holdings, the increases were as follows: Orange +84 %; Vodafone +78 %; OTE +58 %; and RCS&RDS +29 %. Given the insignificant holding in an unpopular band acquired by the new entrant, the auction clearly had the effect of tightening the hold over the market of the three large incumbents (Telecom.paper 2012c). All told, 375 MHz of 15-year spectrum and 110 MHz of short-term spectrum were sold for €682 million, but the complex mix of bandwidths makes it difficult to say whether this was a successful auction by the standards of equivalent countries in the EU.

Subsequently, Orange announced that it would be launching in Bucharest before the year-end using the 1,800 MHz band, whereas Vodafone and OTE opted to await until early in 2013 and 2K Telecom until early 2014 (acting as a wholesaler). 2K Telecom noted that it had been forced to shift from WiMAX to LTE because the former lacked economies of scale as well as an adequate range of smartphones.

In October 2012, the regulator published a draft decision permitting 900 MHz and 1,800 MHz spectrum to be re-farmed for 4G. It was expected to be ratified in November, at which point the decision was extended to allow RCS&RDS to provide 4G using its newly-acquired 900 MHz spectrum as soon as its licence was handed over (Cellular-news 2012g).

In November, Vodafone brought forward its launch, providing coverage in ten cities including Bucharest with a maximum downlink of 75 Mbps rising to 100 Mbps in 2013. Huawei built the network using the 1,800 MHz band (TeleGeography 2012y). For its part, Orange launched in December, also in Bucharest, with a maximum downlink of 75 Mbps via a dongle or tablet (TeleGeography 2012aa). It is worth noting that with the entire country receiving its (maximum) 21.7 Mbps HSPA+ service and most large cities its (maximum) 43.2 Mbps service, Orange's limited new LTE service was barely going to be noticed by consumers. As for OTE, it announced that it would be launching in April with a maximum uplink of 37.5 Mbps and a maximum downlink of 75 Mbps (Telecom.paper 2013e).

In April 2013, Zapp ceased to provide services using CDMA450 technology, thereby freeing up the spectrum, comprising 4.5 MHz paired, for use by interested parties upon request, subject to the proviso that if all requests could not be satisfied then some sort of auction or beauty contest would take place.

5.29 Slovakia

The regulator intended to sell off spectrum in the 800 MHz, 1,800 MHz and 2.6 GHz bands during 2012Q2 on a technology-neutral basis, but this was delayed until 2013Q1. It expressed the view that the incumbents already had enough spectrum to launch 4G, and hence that the sale would be set up so as to encourage new entry. This would best be achieved by setting aside 2G bandwidth for a new entrant and allowing the winner to apply for 2.6 GHz spectrum. The winner(s) would have to launch within 6 months of receiving a licence and provide 25 % population coverage within 1 year, 70 % within 2 years and 98 % by the end of 2017.

The current plan is to arrange an initial single-bid tender for three 800 MHz licences, each of 10 MHz paired—unusually, there is no provision in law to conduct an auction. Since there will be a cap of 15.2 MHz paired per operator in the 1,800 MHz band, and the three incumbents already own that much spectrum, any 1,800 MHz spectrum for sale will necessarily be restricted to new entrants. In practice, although a total of 19.4 MHz paired is available, the licence will be limited to 15.2 MHz paired on grounds of equity.

In the 2.6 GHz band, 70 MHz paired (2,500–2,570 MHz paired with 2,620–2,690 MHz) and 50 MHz unpaired (2,570–2,620 MHz) became available at the end of 2011. A bidder was expected to apply for a minimum of 20 MHz paired, but would also be able to apply for unpaired spectrum in a parallel tender in order to ameliorate interference problems (TeleGeography 2012d).

In July 2012, Telefónica announced that it would shortly be conducting a trial in three villages using the 1,800 MHz band—this is incorrectly described as a launch in Global mobile Suppliers Association (2012). In September, the regulator added that it intended to issue two licences in the 900 MHz band covering the period 2021–2026 as an aid to getting all expiring 2G licences onto the same time-span when renewed.

5.30 Slovenia

In July 2012, Telekom Austria (Si.mobil) launched in Ljubjiana, Brnik and Bled using the 1,800 MHz band (TeleGeography 2012o). In October, the regulator awarded Neuron the right to test LTE in Ljubljana between October 2012 and January 2013, using 10 MHz paired in the 700 MHz band. In November, Telekom Slovenije (Mobitel) announced that it had contracted with Ericsson to roll out a VoLTE-ready LTE network in the 1,800 MHz band. This was duly launched in 27 localities in March 2013 with a view to 40 % population coverage by the year-end (Telecom.paper 2013c).

An auction of 800 MHz band spectrum is pencilled in for 2013. However, it appears that this will be preceded by an auction of additional 1,800 MHz and 2.1 GHz bandwidth in June. On offer are eight 5 MHz paired blocks (1,710–1,720 MHz and 1,755–1,785 MHz paired with 1,805–1,815 MHz and 1,850–1,880 MHz) in the 1,800 MHz band, valid (only) until 3 January 2016, and two 5 MHz paired blocks (1,955–1,965 MHz paired with 2,145–2,155 MHz) in the 2.1 GHz band, valid until 21 September 2021. The coverage requirement is 30 % of the population within 1 year of the licence award. Telekom Austria and Telekom Slovenije will also be given the opportunity to renew their 900 MHz licences at a cost of €5,256 per MHz per month (TeleGeography 2013d). Tusmobil is the third operator qualified to participate in the auction.

5.31 Spain

Telefónica commenced trials pending a series of technology-neutral auctions in 2011. The first of these auctions, in May, resulted in the award (in June) of 5 MHz paired in the 900 MHz band to Orange (which promised to invest $620 million over 3 years on top of the fee of $180 million for a licence lasting until the end of 2030) and in the 1,800 MHz band to TeliaSonera subsidiary Yoigo (which promised to invest $429 million on top of the fee of $60 million) for three licences conferring 5 MHz paired, 5 MHz paired and 4.8 MHz paired respectively (TeleGeography 2011a). Telefónica and Vodafone were excluded from bidding and no interest was forthcoming from new entrants. The new spectrum was earmarked for LTE in rural areas.

Further auctions involving 58 blocks/270 MHz of spectrum in the GSM, digital dividend band (available in 2014) and 2.6 GHz band (available immediately) commenced in June, with operators obliged to commit funds at least equal to their licence payments by 2013. This spectrum was expected to be set aside for LTE but the time scale is as yet unclear. Eleven bidders took part, including most mobile and cable operators, but the opening bids were surprisingly modest and the €1.65 billion raised fell well short of the €2 billion anticipated by the government. Of this €1.65 billion, 98.5 % was bid by three incumbents despite the fact that they were restricted by their spectrum caps: Telefónica (which obtained 10 MHz paired in the 800 MHz band, 5 MHz paired in the 900 MHz band and 20 MHz paired (on a national basis) in the 2.6 GHz band at a cost of €668 million), Orange (which obtained 10 MHz paired in the 800 MHz band and 20 MHz paired (on a national basis) in the 2.6 GHz band at a cost of €437 million) and Vodafone (which obtained 10 MHz paired in the 800 MHz band and 20 MHz paired (composed of a full set of regional blocks) in the 2.6 GHz band for €518 million). The fourth incumbent, TeliaSonera (branded as Yoigo), was unsuccessful as a result of non-participation—the typical outcome for the fourth-largest in Europe so far as the 800 MHz band is concerned but less usual with respect to the 2.6 GHz band (Aetha 2012). Yoigo, which at the time held 14.8 MHz paired and 5 MHz unpaired of 3G spectrum plus 14.8 MHz paired in the 1,800 MHz band, argued that it would take a decade to obtain a return in view of the expected high minimum prices set in the auction, but in any event the prior acquisition of 1,800 MHz spectrum was bound to have affected its decision not to participate. One consequence, naturally, was that the three larger incumbents could each safely bid for 10 MHz paired in the 800 MHz band without driving up the price per MHz.

Other successful bidders comprised regional operators ONO (10 MHz paired/2.6 GHz band/9 regions at a cost of €13.3 million), Jazztel (10 MHz paired/2.6 GHz band/3 regions at a cost of €6 million), Euskaltel (10 MHz paired/2.6 GHz band/1 region at a cost of €2.4 million), R Cable (10 MHz paired/2.6 GHz band/1 region at a cost of €1 million), Telecable (10 MHz paired/2.6 GHz band/1 region at a cost of €0.7 million) and Telecom Castilla La Mancha (10 MHz paired/2.6 GHz band/1 region at a cost of €0.6 million).

Left unsold was 50 MHz of unpaired spectrum in the 2.6 GHz band, 4.8 MHz paired in the 900 MHz band and a regional licence in Extremadura providing 10 MHz paired in the 2.6 GHz band. One factor accounting for this was the existence of an overall spectrum cap imposed across all of the available bands, which meant that operators ignored the bandwidth with the lowest value relative to its price—in this case the unpaired 2.6 GHz spectrum. This was to be re-auctioned, but it was unlikely that there would be any bidders unless existing spectrum caps were to be increased. It is notable that the 30 MHz paired in the 800 MHz band fetched €1.31 billion whereas the 70 MHz paired in the 2.6 GHz band fetched only €172 million, indicating yet again that the latter band is not popular in Europe for LTE.

Vodafone launched a corporate service in Barcelona, Madrid and Málaga, and Telefónica a corporate service in Barcelona and Madrid, both in September 2011. However, Vodafone did not appear to treat this as a commercial launch (see below).

As noted above, LTE is being introduced on a regional as well as national basis, and it was in the Región de Murcia that the Consortium of Advanced Telecommunications (COTA), acting in tandem with Wimax Online to which it will provide wholesale capacity, announced in October 2012 that it would be launching Southern Europe's first regional TD-LTE network in November (Cellular-news 2012f). COTA is reported to have obtained 10 MHz in the 2.6 GHz spectrum for this launch in September 2011 as part of the above auction, paying a seemingly very high price of €157 million ($257 million), but it is not included in the list of regional winners and the price seems excessive.

Meanwhile, in February 2012, Telefónica had unveiled what it called 'the first live experience of the world's smartest 4G network' at the Mobile World Conference. Using the 2.6 GHz band, this would provide up to 100 MHz downstream and between 40 MHz and 60 MHz upstream. Alcatel-Lucent's LightRadio technology would provide much improved indoor coverage and also an initial step towards a real heterogeneous network (HET)—a system whereby conventional radio base stations coexist with 4G metro cells (small base stations incorporating antennas and radio) working on the same frequency and with no interference (TeleGeography 2012c).

Both Orange and TeliaSonera (which had put Yoigo up for sale) set out to launch commercially in July 2013 using the 1,800 MHz band (TeleGeography 2013i). Orange allegedly launched in mid-June in six cities (Telecom.paper 2013h) while TeliaSonera duly launched in Madrid in July providing a maximum downlink of 75 Mbps. They claimed that these would be the first commercial launches in Spain. However, Vodafone evidently got there first by launching in seven cities in late May using the 1,800 MHz and 2.6 GHz bands and claiming to be able to provide a maximum 150 Mbps downlink (Telecom.paper 2013g), and the Orange launch was also stated to have taken place in early July (TeleGeography 2013n). For its part, Telefónica now felt obliged to follow suit even though it preferred to wait for digital dividend spectrum to become available because its 1,800 MHz holding was in near capacity use for 2G. It now intends to launch initially in the 2.6 GHz band in October 2013 although it may sign a roaming agreement with TeliaSonera in the meantime.

5.32 Sweden

The technology-neutral auction in Sweden for spectrum in the 2.6 GHz band was completed in May 2008 and it is worth bearing in mind that at that time there was a shortage of information concerning the future availability and use of the 800 MHz band. Telenor, Tele2 and TeliaSonera each won 20 MHz paired at a cost of €57 million, €59 million and €60 million respectively, while Hutchison-controlled Hi3G Access won 10 MHz paired at a cost of €32 million. The 50 MHz of unpaired spectrum was won by Intel Capital Corp. at a cost of €17 million with a view to launching WiMAX, but was bought by Hi3G Access in December 2010 for an unspecified price. In the run-up to the launch of LTE in Sweden, the incumbents made it clear that their approach would be much more conservative than in the case of 3G. In particular, Telenor and Tele2 announced that they would be pooling their 900 MHz and 2.6 GHz spectrum as Net4Mobility, subject to regulatory approval, while Hi3G Access, which had been one of the first to rollout HSPA+, initially resolved to be a follower in respect of LTE but by late 2010 was talking in terms of a LTE launch in 2011 using both FDD and TDD.

TeliaSonera launched successively in Stockholm—commencing on 13 December 2009 (Telecoms.com 2009)—Visby (June 2010), Gothenburg (August 2010), Malmö and Lund. The intention was to cover 28 locations by the end of 2010 and 228 locations by the end of 2011—for updates see http://www.telia.se/4G. It is of no small interest that whereas the TeliaSonera service was advertised initially as delivering a downlink of 'up to 50 Mbps' using a Samsung modem, it rarely exceeded 10 Mbps at the beginning of 2010 (although the uplink was surprisingly fast at 5 Mbps). This meant that a switch to HSPA outside the service area had little or no effect on download times (Gabriel 2010). For their part, Telenor and Tele2 set out to cover 99 % of the population by the end of 2012 via their jointly-held Net4Mobility network, commencing with commercial launches in Stockholm, Gothenburg and Malmö in November 2010. This coverage was indirectly claimed to have been achieved in March 2013 using the 900 MHz band, with Tele2's 2.6 GHz network achieving 60 % coverage (TeleGeography 2013e). It was subsequently claimed, in July, that the coverage had been achieved using a combined 800/900/1,800/2,600 MHz network with further improvements in capacity due to a doubling of the spectrum in the 900 MHz band allocated for LTE.

There was a further auction of six 25-year licences of 5 MHz paired in the 800 MHz digital dividend band in February/March 2011, analogue transmissions having ceased in October 2007. Although the licences were issued on a technology-neutral basis, they appeared to be ideal for LTE. Hi3G Access (at a cost of SEK431 million), Net4Mobility (at a cost of SEK469 million plus a coverage charge of SEK300 million to pay for homes and businesses lacking broadband) and TeliaSonera (at a cost of SEK854 million) each won 10 MHz paired, the maximum permitted per bidder (PTS 2011). Com Hem and Netett Sverige were unsuccessful bidders. The total amount raised—SEK2,054 million or $325 million—was fairly low for reasons set out in Marshall (2011). Hi3G Access announced in March that it had contracted with ZTE to build the world's first dual-mode LTE TDD/FDD network in both Sweden and Denmark, which appeared to represent its new desire to be a leader rather than, as previously announced, merely a follower. The first stage was launched in December 2011 (Gabriel 2011b). During 2012, further steps were to be taken to upgrade the network which combines FDD spectrum in the 800 MHz and 2.6 GHz bands plus UMTS/900 MHz bands providing HSPA+ with a (maximum) 42 Mbps downlink alongside TD-LTE 2.6 GHz spectrum in the major cities (Cellular-news 2012a).

Technology and service neutral licences providing 35 MHz paired in the 1,800 MHz band, divided into seven blocks of 5 MHz paired, were auctioned in October 2011 with a 25-year life and a start date for usage of 1 January 2013. Existing licensees and new entrants were free to bid for the entire bandwidth on offer. The winners were Net4Mobility (10 MHz paired costing SEK430 million to add to its existing 25 MHz paired) and TeliaSonera (25 MHz paired costing SEK919 million to add to its existing 10 MHz paired) but Hi3G Access came away with nothing (Cellular-news 2011b). As there were only two winners, their respective total spectrum holdings were placed into contiguous spectrum bands—it may be noted that most of the existing licences will expire in 2027 rather than 2037.

However, not everything is proceeding smoothly. In November 2011, a proposed auction of 15 MHz in the 2,010 MHz band was postponed for 1 year pending clarification from the European Commission on harmonisation of the band, and in December only one of 13 county licences in the 3.5 GHz band was sold. For further details of planned auctions see Cellular-news (2011d).

In March 2012, TeliaSonera announced that it had achieved 100,000 subscribers in total on its various LTE networks. However, it admitted that it had launched early, especially in Sweden, mainly for reasons of prestige—it had failed to win a 3G licence in the assignment of 2000—and that subscribers had been put off by a combination of a restricted range of devices, high data prices and limited coverage. By late May, with two-thirds of the population now able to receive LTE, subscriber numbers had risen to 140,000. At the end of June, in comparison, Tele2 claimed to have 70,000 customers but also noted that the great majority of all handsets being sold were LTE-enabled. In February 2013, it noted that it had introduced international LTE roaming between Sweden and Denmark, claiming to be the first operator in Europe to provide such a service.

5.33 Switzerland

The regulator pencilled in an auction for the summer of 2011, with interest to be registered by incumbents and potential new entrants by March. With most GSM licences due to expire in 2013, and 3G licences in 2016, this spectrum was to be offered together with digital dividend spectrum and 4G spectrum—that is, the 800 MHz, 900 MHz, 1,800 MHz, 2.1 GHz and 2.6 GHz bands comprising, in all, 61 blocks (Global Insight 2010). Some of the 900 MHz, 1,800 MHz and 2.1 GHz spectrum could be re-farmed with immediate effect from existing uses and the 2.6 GHz spectrum was also immediately available whereas the 800 MHz blocks would become available in 2013 and the rest during 2015/2017. The coverage requirement for the 800 MHz band, set at 50 % of the population by end-2018, was modest by European standards. The minimum price per block was expected to be high and bidders had to specify the upper limit of spectrum that they wished to acquire in each band. The four incumbents were certain to bid, and Swisscom had already conducted a trial in Grenchen in October 2010 with a view to a launch in

Table 5.8 Auction results. Switzerland. 2012

	800 MHz	900 MHz	1,800 MHz	2.1 GHz	2.6 GHz	$ million
Orange	10 paired	10	50	40	40	170
Sunrise	10 paired	30	40	10 paired	50	528
Swisscom	10 paired	15 paired	30 paired	30 paired	20 paired/45 TDD	395
Total MHz	60	70	150	120	175	
Total $mn						1,093

2011, although a lack of progress with allocating licences caused it to run a number of new trials during 2011/2012.

In the event, an auction took place in late February 2012, but only Orange, Swisscom and Sunrise qualified—the relatively recent licensee, In&Phone, was unsuccessful as was to be expected in the light of experience elsewhere in Europe. As shown in Table 5.8, Orange spent $170 million in return for 10 MHz paired in the 800 MHz band, 10 MHz in the 900 MHz band, 5 lots of 10 MHz in the 1,800 MHz band and 4 lots of 10 MHz in each of the 2.1 GHz and 2.6 GHz bands. Swisscom spent $395 million in return for 10 MHz paired in the 800 MHz band, 15 MHz paired in the 900 MHz band, 30 MHz paired in the 1,800 MHz band, 30 MHz paired in the 2.1 GHz band (FDD) and 20 MHz paired plus 45 MHz unpaired in the 2.6 GHz band. Finally, Sunrise spent $528 million in return for 10 MHz paired in the 800 MHz band, 3 lots of 10 MHz in the 900 MHz band, 4 lots of 10 MHz in the 1,800 MHz band, 10 MHz paired in the 2.1 GHz band and 5 lots of 10 MHz in the 2.6 GHz band (Cellular-news 2012d).

In March 2012, Orange was bought by Apex Partners and re-branded as Matterhorn Mobile. It announced in July that it intended to have a LTE network up-and-running in six cities by the end of 2013 (TeleGeography 2012l), subsequently amended to ten cities in June 2013. For its part, Sunrise announced in July that it would be launching in 11 towns and communities by end-2013, subsequently amended to an initial launch in Zurich, Zug and five winter resorts prior to a full commercial launch in spring 2013. In practice, the launch took place in June (Telecom.paper 2013i) providing up to 100 Mbps to 22 % of its customers (including those of its MVNOs).

Swisscom launched in 26 locations in November 2012, having completed a trial in Berne, Zurich and seven tourist regions, using a combination of the 800 MHz, 1,800 MHz and 2.6 GHz bands. It claimed to provide a maximum downlink of 150 Mbps (Cellular-news 2012k). Population coverage reached 35 % at end-March 2013, at which point it claimed to have 300,000 customers, and was projected to reach 70 % by the year-end. For its part, Matterhorn launched in May with 35 % population coverage—rising to 70 % by the year-end—and a theoretical maximum downlink of 100 Mbps.

References

4G Americas. (2013). *M-Tel launches LTE network in Bulgaria, expands partnership with Ericsson.* Accessed March 29, 2013, from http://www.4gamericas.org

4G-Portal. (2013). *Telecoms operator Telenor launches Hungary's nationwide LTE network with ZTE.* Accessed February 26, 2013, from http://www.4g-portal.com

Adamowski, J. (2013). *Poland raises much more than expected in 1800 MHz auction.* Accessed February 16, 2013 from http://www.policytracker.com

Aetha. (2011a). *Case studies for the award of the 700MHz/800MHz band: Germany.* Accessed November 20, 2011, from http://www.gsma.com/spectrum

Aetha. (2011b). *Case studies for the award of the 700MHz/800MHz band: Finland.* Accessed November 20, 2011, from http://www.gsma.com/spectrum

Aetha. (2012, July). *Spectrum value of 800MHz, 1800MHz and 2.6GHz: A DotEcon and Aetha Report for Ofcom.* DotEcon, London.

Cellular-news. (2011a). *French regulator awards 4G spectrum licenses.* Accessed September 27, 2011, from http://www.cellular-news.com

Cellular-news. (2011b). *TeliaSonera and Net4Mobility buy more radio spectrum in Sweden.* Accessed October 14, 2011, from http://www.cellular-news.com

Cellular-news. (2011c). *Lithuania plans 4G spectrum auction in 2012.* Accessed November 9, 2011, from http://www.cellular-news.com

Cellular-news. (2011d). *Swedish regulators outline radio spectrum auction plans.* Accessed December 20, 2011, from http://www.cellular-news.com

Cellular-news. (2012a). *Hi3G awards contract to ZTE for next stage of dual-mode LTE network.* Accessed March 2, 2012, from http://www.cellular-news.com

Cellular-news. (2012b). *Romanian regulator to hold radio spectrum auction later this year.* Accessed March 20, 2012, from http://www.cellular-news.com

Cellular-news. (2012c). *LTE reaches rural Germany first.* Accessed April 25, 2012, from http://www.cellular-news.com

Cellular-news. (2012d). *Three networks bid for Swiss radio spectrum.* Accessed June 13, 2012, from http://www.cellular-news.com

Cellular-news. (2012e). *Hungarian court blocks launch of fourth mobile network.* Accessed September 17, 2012, from http://www.cellular-news.com

Cellular-news. (2012f). *Nokia Siemens Networks brings TD-LTE to Spain.* Accessed October 19, 2012, from http://www.cellular-news.com

Cellular-news. (2012g). *Romania consults on LTE services in GSM spectrum bands.* Accessed October 29, 2012, from http://www.cellular-news.com

Cellular-news. (2012h). *Ireland raises US$1 billion from radio spectrum auction.* Accessed November 16, 2012, from http://www.cellular-news.com

Cellular-news. (2012i). *Norwegian radio spectrum auction ends after just one round.* Accessed November 19, 2012, from http://www.cellular-news.com

Cellular-news. (2012j). *Over half of EMT customers now selecting LTE services.* Accessed November 25, 2012, from http://www.cellular-news.com

Cellular-news. (2012k). *Swisscom launches its LTE network in Switzerland.* Accessed November 29, 2012, from http://www.cellular-news.com

Cellular-news. (2013a). *Hungarian court backs MNOs in case against state-owned firm.* Accessed February 28, 2013, from http://www.cellular-news.com

Cellular-news. (2013b). *Czech Republic cancels mobile auction after bids go too high.* Accessed March 11, 2013, from http://www.cellular-news.com

Cellular-news. (2013c). *Samsung wins Irish LTE deployment contract.* Accessed April 12, 2013, from http://www.cellular-news.com

ClearMobitel. (2011). *Press Release – September 26th 2011.* Accessed October 31, 2011, from http://www.clearmobitel.com

Donegan, M. (2010). *Deutsche Telekom reveals next LTE move.* Accessed November 29, 2010, from http://www.lightreading.com

Erhvervsstyrelsen. (2010). *Results of the auction.* Accessed May 21, 2010, from http://www. Erhvervsstyrelsen.dk

Gabriel, C. (2010). *In Europe's 2.6 GHz auctions, prices depressed as EU looks for 800 MHz harmony.* Accessed September 30, 2010, from http://www.rethink-wireless.com

Gabriel, C. (2011a). *Vodafone dumps DSL for LTE in Germany.* Accessed August 24, 2011, from http://www.rethink-wireless.com

Gabriel, C. (2011b). *Hi3G rolls out dual-mode LTE in Sweden.* Accessed December 18, 2011, from http://www.rethink-wireless.com

Gabriel, C. (2012a). *3 Italia to go live with LTE this year.* Accessed March 13, 2012, from http://www.rethink-wireless.com

Gabriel, C. (2012b). *Dutch auction delivers record sums.* Accessed December 17, 2012, from http://www.rethink-wireless.com

Global Insight. (2010). *Swiss regulator plans digital dividend auction for H1 2011.* Accessed December 15, 2010, from https://www.communicationsdirectnews.com

Global mobile Suppliers Association. (2012). *Evolution to LTE report.* Accessed November 27, 2012, from http://www.gsacom.com

Global View Partners. (2009, December). *The 2.6 GHz spectrum band, report.* Prepared for the GSM Association.

Globaltelecomsbusiness. (2012). *Belgacom to launch Luxembourg LTE.* Accessed September 18, 2012, from http://www.globaltelecomsbusiness.com

Hibberd, M. (2012). *Telenor launches LTE in 11 Norwegian cities.* Accessed October 10, 2012, from http://www.telecoms.com

Jóhannsson, A. (2013). *Auction of spectrum in Iceland 800 MHz and 1800 MHz.* Accessed February 10, 2013, from http://www.cullen-international.com

Lennighan, M. (2013). *TeliaSonera wins Estonia 800-MHz spectrum, more airwaves to be offered.* Accessed May 31, 2013, from http://www.totaltele.com

LteWorld. (2012). *TDC, Telia-Telenor win 800MHz LTE frequency auction in Denmark.* Accessed July 11, 2012, from http://www.lteworld.org

Marshall, P. (2011). *Sweden's 800 MHz spectrum auction sees lukewarm response as service providers accelerate their efforts towards 4G.* Accessed March 9, 2011, from http://www. 4gtrends.com

PTS. (2011). *Auction concluded – total amount raised SEK 2,054,000,000.* Accessed April 5, 2011, from http://www.pts.se

Pyramid Research. (2010). *German spectrum auction to pay 'digital dividends'.* Accessed May 11, 2010, from http://www.pyramidresearch.com

Rayal, F. (2011). *Analysis of the 2.6 GHz auction in Belgium.* Accessed December 1, 2011, from http://www.frankrayal.com

Reuters. (2013). *Update 3-Telecom Italia delays decision on Hutchison deal talks.* Accessed May 8, 2013, from http://www.reuters.com

Sahota, D. (2012). *O_2 launches LTE in Czech Republic.* Accessed June 20, 2012, from http://www. telecoms.com

Sahota, D. (2013). *Bouygues Telecom to launch LTE in the next few weeks.* Accessed April 4, 2013, from http://www.telecoms.com

SeeNews. (2012). *Telenor Montenegro launches 4G services in four cities.* Accessed March 24, 2013, from http://www.seenews.com

Telecom.paper. (2012a). *Tele2 NL launches LTE for tablets, laptops.* Accessed May 9, 2012, from http://www.telecompaper.com

Telecom.paper. (2012b). *TDC, Telia-Telenor win Danish LTE frequency auction.* Accessed June 27, 2012, from http://www.telecompaper.com

Telecom.paper. (2012c). *Romania raises EUR 682 million in spectrum auction.* Accessed September 25, 2012, from http://www.telecompaper.com

Telecom.paper. (2012d). *SFR brings forward LTE launch to 28 November*. Accessed September 25, 2012, from http://www.telecompaper.com

Telecom.paper. (2012e). *3 Denmark unveils 40GB pack at LTE launch*. Accessed September 25, 2012, from http://www.telecompaper.com

Telecom.paper. (2012f). *Telecom Italia confirms launch of new LTE services*. Accessed October 16, 2012, from http://www.telecompaper.com

Telecom.paper. (2012g). *Polkomtel offers plus LTE internet at 150 Mbps*. Accessed October 25, 2012, from http://www.telecompaper.com

Telecom.paper. (2012h). *Vodafone brings 4G network to Rome, Milan*. Accessed October 30, 2012, from http://www.telecompaper.com

Telecom.paper. (2012i). *Belgacom launches LTE services in 8 cities*. Accessed November 5, 2012, from http://www.telecompaper.com

Telecom.paper. (2012j). *Cosmote launches LTE network in Athens, Thessaloniki*. Accessed November 19, 2012, from http://www.telecompaper.com

Telecom.paper. (2012k). *Dutch multiband spectrum auction ends with four winners*. Accessed December 14, 2012, from http://www.telecompaper.com

Telecom.paper. (2012l). *Hrvatski Telekom expands LTE to rural areas*. Accessed December 27, 2012, from http://www.telecompaper.com

Telecom.paper. (2013a). *KPN to launch 4G services in early February*. Accessed January 15, 2013, from http://www.telecompaper.com

Telecom.paper. (2013b). *French government agrees to refarm 1800MHz for 4G*. Accessed February 25, 2013, from http://www.telecompaper.com

Telecom.paper. (2013c). *Telekom Slovenije launches LTE services*. Accessed March 22, 2013, from http://www.telecompaper.com

Telecom.paper. (2013d). *Nova launches LTE across Reykjavik*. Accessed April 8, 2013, from http://www.telecompaper.com

Telecom.paper. (2013e). *Cosmote Romania commercially launches LTE services*. Accessed April 30, 2013, from http://www.telecompaper.com

Telecom.paper. (2013f). *Macedonia opens tender for mobile spectrum licences*. Accessed May 14, 2013, from http://www.telecompaper.com

Telecom.paper. (2013g). *Vodafone Spain to launch LTE network on 29 May*. Accessed May 28, 2013, from http://www.telecompaper.com

Telecom.paper. (2013h). *Orange launches 4G services in Spain*. Accessed June 19, 2013, from http://www.telecompaper.com

Telecom.paper. (2013i). *Sunrise launches commercial LTE services*. Accessed June 20, 2013, from http://www.telecompaper.com

Telecoms.com. (2009). *TeliaSonera launches commercial LTE in Stockholm and Oslo*. Accessed December 14, 2009, from http://www.telecoms.com

TeleGeography. (2010a). *Telestyrelsen awards 900MHz, 1800MHz licences to Hi3G*. Accessed September 15, 2010, from http://www.telegeography.com

TeleGeography. (2010b). *ETSA announces winners of 4G spectrum auction*. Accessed December 18, 2010, from http://www.telegeography.com

TeleGeography. (2011a). *Yoigo, Orange win tech neutral spectrum*. Accessed June 14, 2011, from http://www.telegeography.com

TeleGeography. (2011b). *Aero2 and Huawei to launch converged LTE network*. Accessed September 20, 2011, from http://www.telegeography.com

TeleGeography. (2011c). *Three telcos pay EUR380.5m for 900MHz, 1800MHz frequencies*. Accessed November 15, 2011, from http://www.telegeography.com

TeleGeography. (2011d). *France raises EUR2.64bn in 800MHz auction*. Accessed December 23, 2011, from http://www.telegeography.com

TeleGeography. (2012a). *Latvian quartet bag 2600MHz spectrum in PUC auction*. Accessed January 9, 2012, from http://www.telegeography.com

TeleGeography. (2012b). *NMHH clears four to bid for 900MHz spectrum.* Accessed January 9, 2012, from http://www.telegeography.com

TeleGeography. (2012c). *Telefonica unveils 4G trial network at Mobile World Congress.* Accessed February 27, 2012, from http://www.telegeography.com

TeleGeography. (2012d). *Regulator concludes 4G licensing evaluation; aiming for fourth operator.* Accessed March 8, 2012, from http://www.telegeography.com

TeleGeography. (2012e). *Comreg's auction of 4G spectrum 'imminent', but telcos' frustration grows.* Accessed March 9, 2012, from http://www.telegeography.com

TeleGeography. (2012f). *Ireland inches closer to 4G wireless spectrum auction.* Accessed March 21, 2012, from http://www.telegeography.com

TeleGeography. (2012g). *VIPnet launches LTE: Second or first in Croatia?* Accessed March 26, 2012, from http://www.telegeography.com

TeleGeography. (2012h). *Milmex set to launch WiMAX later this year.* Accessed May 1, 2012, from http://www.telegeography.com

TeleGeography. (2012i). *PT confirms TMN LTE is now available to 80% of population.* Accessed May 5, 2012, from http://www.telegeography.com

TeleGeography. (2012j). *Tango makes a song and dance over LTE licence.* Accessed June 19, 2012, from http://www.telegeography.com

TeleGeography. (2012k). *E-Plus can use 1800MHz for mobile broadband.* Accessed June 29, 2012, from http://www.telegeography.com

TeleGeography. (2012l). *Orange sketches plans for 2013 LTE rollout; earmarks CHF700m.* Accessed July 2, 2012, from http://www.telegeography.com

TeleGeography. (2012m). *3 outlines 4G plans.* Accessed July 10, 2012, from http://www.telegeography.com

TeleGeography. (2012n). *CTU confirms final conditions for mobile auction.* Accessed July 11, 2012, from http://www.telegeography.com

TeleGeography. (2012o). *Si.Mobil launches LTE in three cities.* Accessed July 13, 2012, from http://www.telegeography.com

TeleGeography. (2012p). *UKE launches tender for five blocks of 1800MHz spectrum.* Accessed August 21, 2012, from http://www.telegeography.com

TeleGeography. (2012q). *TD boosts LTE speed to 100Mbps for smartphone users.* Accessed August 31, 2012, from http://www.telegeography.com

TeleGeography. (2012r). *AT confirms five in frame for Dutch mobile spectrum auction.* Accessed September 7, 2012, from http://www.telegeography.com

TeleGeography. (2012s). *EU extends H3G, Orange decision to 30 November; T-Mobile complains about regulatory conditions.* Accessed September 17, 2012, from http://www.telegeography.com

TeleGeography. (2012t). *Orange unveils 29 October LTE launch date.* Accessed September 28, 2012, from http://www.telegeography.com

TeleGeography. (2012u). *FICORA seeks comments on 800MHz auction and licensing.* Accessed October 26, 2012, from http://www.telegeography.com

TeleGeography. (2012v). *Mobile auction launches today.* Accessed October 31, 2012, from http://www.telegeography.com

TeleGeography. (2012w). *3 pips TIM with small-scale LTE launch in Acuto.* Accessed November 6, 2012, from http://www.telegeography.com

TeleGeography. (2012x). *Orange launches LTE in four cities.* Accessed November 23, 2012, from http://www.telegeography.com

TeleGeography. (2012y). *Vodafone Romania launches commercial LTE services in ten cities.* Accessed November 23, 2012, from http://www.telegeography.com

TeleGeography. (2012z). *Tele2 Estonia launches LTE.* Accessed November 28, 2012, from http://www.telegeography.com

TeleGeography. (2013a). *Free mobile bags new 900MHz spectrum, as watchdog opposes mooted merger with SFR.* Accessed January 3, 2013, from http://www.telegeography.com

TeleGeography. (2013b). *Bulsatcom becomes Bulgaria's fourth mobile operator*. Accessed January 15, 2013, from http://www.telegeography.com

TeleGeography. (2013c). *Bouygues may have to wait for regulatory approval*. Accessed January 17, 2013, from http://www.telegeography.com

TeleGeography. (2013d). *Slovenia invites bids for 1800MHz, 2100MHz licences; settles 900MHz renewal*. Accessed March 18, 2013, from http://www.telegeography.com

TeleGeography. (2013e). *Tele2 Sweden reaches 99% 4G coverage?* Accessed March 19, 2013, from http://www.telegeography.com

TeleGeography. (2013f). *Tele2 switches on LTE network in five cities*. Accessed March 19, 2013, from http://www.telegeography.com

TeleGeography. (2013g). *TKK unveils 'combinatory clock auction', T-Mobile forced to hire theoretician to understand rules*. Accessed March 19, 2013, from http://www.telegeography.com

TeleGeography. (2013h). *T-Mobile sees frequency fracas referred to European Court of Justice*. Accessed May 9, 2013, from http://www.telegeography.com

TeleGeography. (2013i). *Orange Espana to launch ahead of Yoigo as it reveals 8 July date for LTE switch-on*. Accessed May 13, 2013, from http://www.telegeography.com

TeleGeography. (2013j). *AEC launches 4G tender for 790MHz-862MHz, 1710MHz-1880MHz blocks*. Accessed May 14, 2013, from http://www.telegeography.com

TeleGeography. (2013k). *Free 4G access for Vodafone users until end-October*. Accessed June 13, 2013, from http://www.telegeography.com

TeleGeography. (2013l). *Cyprus publishes tender documents for the 900MHz/1800MHz/2100MHz bands auction*. Accessed July 4, 2013, from http://www.telegeography.com

TeleGeography. (2013m). *BIPT publishes details of upcoming 800MHz spectrum auction*. Accessed July 4, 2013, from http://www.telegeography.com

TeleGeography. (2013n). *Orange inaugurates 4G network in six cities*. Accessed July 9, 2013, from http://www.telegeography.com

TeleGeography. (2013o). *VIP Operator acquires LTE spectrum*. Accessed July 11, 2013, from http://www.telegeography.com

TeleGeography. (2013p). *CRC launches auction for 2500MHz-2690MHz bands*. Accessed July 15, 2013, from http://www.telegeography.com

TeleGeography. (2013q). *Cyprus publishes tender documents for 2600MHz bands auction*. Accessed July 15, 2013, from http://www.telegeography.com

TeleGeography. (2012aa). *Orange Romania launches LTE in Bucharest*. Accessed December 13, 2012, from http://www.telegeography.com

TeleGeography. (2012ab). *EC clears Hutchison Whampoa's plan to acquire Orange Austria*. Accessed December 13, 2012, from http://www.telegeography.com

TeleGeography. (2012ac). *T-Mobile tests LTE-Advanced technology in Vienna*. Accessed December 18, 2012, from http://www.telegeography.com

Telenor. (2012). *Network development at Telenor goes on: 84 Mbps maximum nominal download speed now available in 60 towns*. Accessed November 27, 2012, from http://www.telenor.hu

Total Telecom. (2012). *Czech Republic to auction 4G spectrum*. Accessed March 21, 2012, from http://www.totaltele.com

Total Telecom. (2013). *Germany plans spectrum auction in 2014 - report*. Accessed June 25, 2013, from http://www.totaltele.com

Ventura Team. (2010). *Has Hi3G played a shrewd hand in the recent Danish auctions? Unpaired vs. paired spectrum analysis*. Accessed June 29, 2010, from http://www.unstrung.com

WirelessMoves. (2012). *French LTE 800 auction results: Background and questions*. Accessed July 17, 2010, from http://mobilesociety.typepad.com

Wood, N. (2012). *Clearwire Belgium renamed b lite*. Accessed September 17, 2012, from http://www.totaltele.com

Asia-Pacific

6

6.1 Afghanistan

In February 2013, Etisalat issued a press release which implied that it was in a position to provide 4G services. The regulator promptly intervened to deny that any 4G licences had been issued, and asked Etisalat to withdraw its claims (Cellular-news 2013e).

6.2 Australia

A preliminary issue that merits some comment is that when spectrum was originally assigned it was done for the purposes of voice transmission. Hence, for example, the 1,800 MHz band was divided into regional licences often with small contiguous lots meaning that the ideal contiguous bandwidth for LTE of 20 MHz was not available even regionally, let alone on a nationwide basis, a situation compounded by the holdings in the band in the hands of the state railway operators. Telstra, holding 1,710–1,725 MHz paired with 1,805–1,820 MHz in the majority of states, was initially best positioned but even then it found itself with non-contiguous spectrum in the crucial markets of Canberra, Melbourne and Sydney. Although this could be remedied without difficulty via spectrum swaps in a number of cases such as Melbourne, it is likely to continue to be a factor holding back the speedy roll-out of LTE. At the present time, because Vodafone has negotiated a number of such swaps, it is able to provide the fastest service in areas where it now has 20 MHz of contiguous spectrum.

In May 2011, the regulator announced that in November it would be re-allocating two 45 MHz blocks of digital dividend spectrum in the 703–748 MHz and 758–803 MHz bands on a nationwide basis for LTE use, divided into nine lots of 5 MHz paired—for the background see Aetha (2011). Australia has opted for the APT band plan. The spectrum would be auctioned in November 2012—postponed until April 2013—and become available on 31 December 2013 with a 15-year

P. Curwen and J. Whalley, *Fourth Generation Mobile Communication,*
Management for Professionals, DOI 10.1007/978-3-319-02210-9_6,
© Springer International Publishing Switzerland 2013

time-span. The regulator would simultaneously be re-allocating two blocks in the 2,500–2,570 MHz and 2,620–2,690 MHz bands with availability other than in Perth by end-2013. 14 lots of 5 MHz paired would be made available, but on a regional basis with the facility to combine these into a national licence (Cellular-news 2012a). Unusually, a Combinatorial Clock Auction format would be used to allow bidders to buy packages of multiple spectrum bands in a single auction.

In the meantime, Telstra anticipated a limited launch commencing in May 2011 using 1,800 MHz spectrum freed up through the transfer of customers to its Next G network (using the 850 MHz band)—the official launch took place in December. Vodafone Hutchison Australia (VHA), formed in June 2009 and half-owned by Vodafone and Hutchison, intended to follow suit in 2013 using the 1,800 MHz band (Cellular-news 2012c) in conjunction with the Australasian Railway Association. For its part, Optus pencilled in a launch using the 1,800 MHz band in April 2012, and in September 2011 it was the first operator to be awarded a licence to trial LTE using digital dividend spectrum (Cellular-news 2011). The trial commenced in early November and was completed in March 2012 having achieved a maximum downlink of 70 Mbps and uplink of 32 Mbps.

In February 2012, Optus announced that it had conditionally agreed to buy WiMAX operator Vivid Wireless, thereby acquiring 98 MHz of spectrum in the 2.3 GHz band which it would convert to TD-LTE and integrate with its other LTE network. It launched in Sydney, Newcastle and Perth using the 1,800 MHz band in July 2012—restricted to corporate customers with a service for the general public following on in September—and announced that its wholly-owned subsidiary Virgin Mobile Australia, acting as a MVNO over its network, would also offer LTE in September 2012. Optus LTE services are currently also provided by iiNet (since October 2012) and Exetel (since February 2013).

Also in February 2012, the communications minister announced that a cap of 20 MHz paired in the 700 MHz band and of 40 MHz paired in the 2.6 GHz band would apply to each bidder in the forthcoming auction (TeleGeography 2012c). For its part, VHA announced that it would be concentrating upon the roll-out of '4G-like' HSPA+.

In May, Optus and VHA announced that, as from April 2013, they would be sharing base stations for their 3G and 4G networks (subject to regulatory approval) with a further 500 stations to be added to the joint network (TeleGeography 2012k).

It is also worth noting that, as the party responsible for the creation of a national broadband network, National Broadband Network (NBN Co) launched a fixed-wire TD-LTE network using the 2.3 GHz band in April 2012. Operating at relatively slow speeds in rural areas, NBN Co provides a wholesale service. EnergyAustralia's Ausgrid also has ambitions to develop a LTE network.

Telstra, which has announced its intention to cover 67 % of the population with LTE by mid-2013, also has ambitions to be amongst the first operators to provide LTE-Advanced using re-farmed spectrum in the 900 MHz and 1,800 MHz bands (TeleGeography 2012m). It is testing, among other things, LTE-A Carrier Aggregation. In January 2013, Telstra signed an international roaming agreement with Hong Kong's CSL. In May, it claimed to have sold 2.1 million devices of which

1.4 million were handsets—the longest that any operator had taken to cross the two million threshold thus far although few had achieved this.

The April 2013 auction became mired in controversy when, as a result of VHA hinting that it would not participate if the floor price was set too high, the Minister for Communications stepped in to claim responsibility for spectrum pricing (Telecom.paper 2012n). In December, both VHA and Optus reacted negatively to the reserve price of A$1.36 ($1.43) per MHz per pop set by the Minister for 700 MHz band spectrum which far exceeded anything previously achieved in digital dividend spectrum sales—the equivalent reserve for the 2.6 GHz spectrum was a mere $0.03 per MHz per pop. Claiming that it was not short of spectrum overall, VHA stated that it was conducting trials in Sydney using a 10 MHz carrier in the 1,800 MHz band, half the bandwidth that it intended to use when launching commercially. It promised a downlink of between 2 Mbps and 40 Mbps upon launch in June (TeleGeography 2013a) which took place in seven cities.

The results of the auction in early May were as follows:
- Telstra: 20 MHz paired 700 MHz; 40 MHz paired 2.6 GHz; $1.26 billion
- Optus: 10 MHz paired 700 MHz; 20 MHz paired 2.6 GHz; $630 million
- TPG Internet: 10 MHz paired 2.6 GHz; $13.1 million

The new entrant, TPG Internet, is currently a reseller using the Optus network. Altogether, the auction raised $1.90 billion but 30 MHz of spectrum in the 700 MHz band was left unsold. The 700 MHz band licences will become operational in January 2015 while the 2.6 GHz band licences will become operational in October 2014 other than in the Perth metro area and regional Western Australia where they will commence in February 2016 (Taylor 2013).

After the auction was completed, the regulator was accused of setting too high a reserve price for the 700 MHz spectrum but responded that VHA had already expressed its intention not to bid before the reserve price had been fixed and had also declined to bid for spectrum in the 2.6 GHz band. He was also at pains to point out that a new entrant had emerged. A debate is currently waging over the future of the unsold spectrum and the share that might be handed over to public safety bodies.

In June 2013, Optus launched TD-LTE in Canberra using the 2.3 GHz band, but as yet there are no dual-band handsets available. VHA (which began using the Vodafone brand name in August) launched in July providing a downlink of between 10 Mbps and 42 Mbps and an uplink of between 2 Mbps and 21 Mbps (Porter 2013)—to which Telstra responded by threatening to sue Vodafone for claiming incorrectly that its network was the faster. In August, Vodafone began offering the Huawei MediaPad 10 Link 4G, but this appeared to be a touch optimistic given that it is a Category 4 device capable of handling a downlink of 150 Mbps.

6.3 Bangladesh

In late March 2012, the regulator announced that, as part of the procedure whereby 15-year 3G licences were to be awarded—of which six were to be offered, comprising one for state-owned BTCL and two with four for other incumbents and one for a new entrant (implying that at least one of the six 2G incumbents would not get a licence)—the winning bidders would be able to upgrade to 4G licences at no extra cost (Telecom.paper 2012e). A total of 50 MHz paired in the 2.1 GHz band would be available with the reserve price set at $50 million per MHz paired (TeleGeography 2012p). A successful new entrant would be able to apply for a GSM licence at a discounted price at a later date.

However, the 3G auction was subsequently delayed until 2013 due to ongoing disputes over the renewal of four 2G licences. As a result, the regulator announced that this would, in turn, delay the auction of digital dividend spectrum which in any event required some re-farming. In July 2006, 6 MHz paired had been awarded at no charge to an ISP, AlwaysOn Network Bangladesh, but this needed to be shifted within the band to expedite the auction.

At the end of January 2013, the regulator announced that the 50 MHz paired of 3G spectrum would now be divided into 10 blocks of 5 MHz paired in order to allow in principle for all six incumbents and/or more than one new entrant to obtain a licence. By this point in time BTCL (Teletalk) had already launched (in October 2012) so it would be required to pay in accordance with the outcome of the auction (Cellular-news 2013a). The auction date was subsequently set as 24 June—but delayed again until July due to uncertainty as to the rate of VAT to be levied on licence payments—and the reserve price at $20 million per MHz paired, with individual bids capped at two blocks unless there were insufficient bids, in which case three blocks could become available. A further delay until September was then announced.

In July, the regulator stated that any bidders acquiring two or more blocks would be allowed to install LTE. Furthermore, any bidder acquiring more than two blocks would receive the additional blocks at a discounted price. As BTCL had already been awarded two blocks, four further licences, each of 10 MHz, would be available to private companies although, as noted, a minimum bid of one block could be tabled. One licence would be reserved for a new entrant and, if none was forthcoming, the licence would become available for an incumbent.

The position in relation to WiMAX became mired in controversy when a decision made by the regulator in August 2011 to allocate without charge a block of spectrum in the 800 MHz band (806–816 MHz paired with 847–857 MHz) to New Generation Graphics Limited (NGGL) with a view to the provision of WiMAX under the Ollo brand was challenged in the High Court by WiMAX operator Banglalion. At the time Ollo was available only in Dhaka City even though the 800 MHz licence had been issued in order to improve rural connectivity, but Ollo claimed to be using fixed-wireless in the 3.5 GHz band. For its part, the regulator claimed that it had no powers to refuse a request by the Prime Minister's Office under the Access to Information (A2I) project to issue 800 MHz spectrum to an ISP, and that it had previously issued spectrum to other ISPs. Banglalion pointed out in its petition that a joint venture had subsequently been formed between NGGL, Bangladesh Internet Exchange Limited (BIEL) and its parent Russia's Multinet, that Ollo was using unlicensed equipment and that Ollo was preparing for a launch of LTE even though this had yet to be approved. The Court duly stayed the allocation in May 2013 (TeleGeography 2013g). Subsequently, Ollo applied for an unissued licence in the 2,330–2,365 MHz band—those successfully issued in 2008 comprised 2,365–2,400 MHz awarded to Qubee and 2,585–2,620 MHz awarded to Banglalion.

6.4 Bhutan

In February 2013, Ericsson announced that it was rolling out a network for Bhutan Telecom using the 1,800 MHz band, with a view to an initial launch in Thimphu during 2013 providing up to a 40 Mbps downlink (Cellular-news 2013f).

6.5 Brunei

In November 2012, the regulator announced that the 1,800 MHz band was the only one suitable for immediate deployment for LTE, consisting of 75 MHz paired split into 5 MHz paired blocks. Two 15-year licences would be awarded via a beauty contest with licensees required to achieve nationwide coverage within 3 years of receiving a licence (TeleGeography 2012u).

6.6 China

China Mobile has conducted a series of TD-LTE trials—the first phase consisting of a six-city trial was completed in October 2011 while the second phase lasted until July 2012 and involved 20,000 base stations—which it intends to increase tenfold by the end of 2013 (TeleGeography 2012o) so what is being referred to as trials is not that dissimilar to what is being referred to as a rolling commercial launch in other countries. The approach involves an overlay on top of the existing TD-SCDMA network combined with Wi-Fi. In contrast, China Telecom is conducting trials using paired spectrum. The surprise package is a joint venture involving US-based VelaTel Global Communications—which is rolling out a jointly-owned TD-LTE network built by strategic partner ZTE—and New Generation Special Network Communication Technology (NGSN), a value added service (VAS) licensee. NGSN is to play the role of anchor tenant and as a wholesaler of spare capacity (TeleGeography 2011c).

In March 2012, it was announced that 220,000 TD-SCDMA (3G) base stations had been constructed to date, and that no attempt would be made to issue 4G licences until this total had been raised to 400,000 (provided by a wider range of vendors), which would take a further 2–3 years. This came as a blow to China Mobile which had been forced to adopt the unpopular TD-SCDMA standard (rather than W-CDMA) for 3G and as a result had been pushing to launch TD-LTE before the end of 2012. While it could arguably use re-farmed 2G/3G spectrum—it owns spectrum in the 1,900 MHz, 2.0 GHz, 2.3 GHz and 2.6 GHz bands—this would provide little more than a stop-gap solution. For their part, neither China Unicom nor China Telecom has been particularly forthcoming about its intentions.

Mixed messages have been appearing as to when licences will officially be awarded, but it is looking increasingly likely that this will happen before the end of 2013. One recent pronouncement indicated that 98 MHz of spectrum in the 1,477–1,515 MHz band would be made available for TD-LTE, although it is unclear how much of this would be allocated to China Mobile, to be added to the 190 MHz previously set aside in the 2.6 GHz band.

The Ministry of Information & IT (MIIT) is responsible for choosing vendors. So far, 11 have put their names forward and taken part in the second phase of TD-LTE trials involving the MIIT and China Mobile. As of May 2012, only ZTE had satisfied the requirements for chipsets, terminals and systems. The ZTE ZX297502 chipset was the first one of its kind—combining TD-LTE, TD-SCDMA, FDD-LTE and GSM—to pass muster (Cellular-news 2012b). In September, Sequans' LTE technology was also approved.

In February 2013, China Mobile announced that it was well on its way towards meeting its target of 200,000 TD-LTE base stations by the year-end—it put out a tender for an additional 207,000 base stations in June 2013. On that basis, China Mobile expected to account for all of the roughly one million subscribers to LTE signing up during 2013 assuming a launch in August in 344 cities. However, the Ministry of Industry and Information Technology (MIIT) is allegedly on the point of issuing TD-LTE licences to all three incumbents while requiring them to apply for FDD licences. Probably in response, China Telecom stated that it would now be using paired spectrum other than in compact urban areas where TD-LTE would be used.

6.7 Fiji

In November 2012, Unwired Fiji (in partnership with Digicel) claimed to have launched '4G Broadband'. The technology involved was WiMAX, but as only 8 Mbps at best was available the claims of 'super-fast' connectivity appeared both a tad optimistic and hard to equate with the common understanding of 4G (Tawakilagi 2012). At the end of May 2013, the government announced that it would be holding a LTE auction 1 month later without specifying either licence conditions or reserve prices. It subsequently added that spectrum in the 700 MHz, 800 MHz and 1,800 MHz bands would be available, divided into 20 blocks of 5 MHz apiece, with bids to be submitted by mid-June. The blocks were divided into 'Standard' and 'Special Coverage Requirement' with different reserve prices set accordingly.

A cap of 30 MHz was imposed on individual bidders. Digicel, Vodafone and Telecom Fiji were the registered bidders when the auction began in late July, with Vodafone initially acquiring 30 MHz in the 1,800 MHz band at a cost of $460,000, Digicel acquiring 15 MHz in the 700 MHz band at a cost of $680,000 and Telecom Fiji acquiring 15 MHz at a cost of $816,000 (TeleGeography 2013m). Telecom Fiji acquired a further 15 MHz and Digicel a further 5 MHz, but reports are contradictory about both spectrum bands and prices other than that the total raised was a modest $5 million.

6.8 Guam

IT&E launched in August 2012 (TeleGeography 2012q), DoCoMo in October 2012 (Telecom.paper 2012k) and iConnect (branded as LTE True 4G and providing a maximum downlink of 21 Mbps and uplink of 5 Mbps) in March 2013 (TeleGeography 2013c), all using the 700 MHz band.

6.9 Hong Kong

Hong Kong is auctioning several lots of spectrum that can potentially be used for LTE. In October 2008, the government announced the initial release of spectrum in the 2.3 GHz and 2.6 GHz bands—in total, 195 MHz in 12 bandwidths varying from 5 MHz to 30 MHz would be auctioned, commencing in January 2009, with open entry and any combination of bands available subject to an individual bidder cap of 30 MHz. A simultaneous ascending auction would be used for all bands together, with each band going to the highest bidder subject to a minimum of HK$25 million per 5 MHz block. Services would have to be provided within 5 years. Successful bidders in January 2009 were CSL New World, China Mobile People's and Genius Brand (Hutchison/PCCW) which each won 15 MHz paired in the 2.6 GHz band at a total cost of $198 million. SmarTone and Hong Kong Broadband Network were unsuccessful. 15 MHz in the 2.6 GHz band and all of the 2.3 GHz band spectrum was left unassigned.

During 2010, CSL claimed that it had run a successful trial and would be in a position to launch before the year-end. It would launch a dual-band network using re-farmed 1,800 MHz spectrum together with its newly-acquired spectrum. In practice, the November launch used only the 1,800 MHz band and was only for the benefit of selected corporate customers, and the consumer launch in May 2011 was restricted initially to HSPA+ customers followed by a full launch in August. The 1,800/2,600 MHz dual-band service was not launched until August 2012. For its part, SmarTone specified that it would initially restrict itself to its existing 1,800 MHz spectrum.

In December 2010, the regulator announced plans to auction spectrum in the 850 MHz, 900 MHz and 2 GHz bands in February 2011. The available bands were 'A' (832.5–837.5 MHz paired with 877.5–882.5 MHz, 'B' (885–890 MHz paired with 930–935 MHz), 'C1' (2,010.0–2,014.8 MHz) and 'C2' (2,014.8–2,019.7 MHz). The auction was open to all-comers who could bid for any or all of the bands with licences lasting for 15 years. Winners would be required to achieve 50 % population coverage within 5 years. Band 'B' came with conditions attached such as the need to share spectrum with the railways (Cellular-news 2010). In April 2011, the 'A' band licence was awarded to SmarTone for $112 million (allocated by SmarTone for use with HSPA+) and the 'B' band licence to Hutchison Telephone for $132 million (Globaltelecomsbusiness 2011).

In November 2011, the regulator announced details of a 2.3 GHz auction, commencing in February 2012, comprising three 15-year licences, each of 30 MHz, open to all-comers. The coverage requirement if a mobile service was to be provided was 50 % of the population. The applicants were China Mobile, Hong Kong Telecommunications, Hutchison Telephone and 21 ViaNet Group (formerly AsiaCloud). The winners were China Mobile (at a cost of $21.9 million), Hutchison and new entrant 21 ViaNet (each paying $19.4 million) (Telecom.paper 2012a). Hutchison (trading as 3 Hong Kong) subsequently announced that it intended to launch a FDD/TDD dual-mode service using this spectrum. For its part, China Mobile launched in April 2012 using the 2.6 GHz band (TeleGeography 2012h) and set out to add TD-LTE using the 2,300 MHz band in 2012Q4 which it duly achieved in December albeit serving only selected commercial customers (TeleGeography 2012v).

In January 2012, the regulator undertook consultations on the release via an auction of 25 MHz paired—split up into five 15-year technology-neutral licences—in the 2.5/2.6 GHz band (2,515–2,540 MHz paired with 2,635–2,660 MHz), including the 15 MHz paired left unsold in 2009. There would be no spectrum cap but a 5-year minimum coverage condition would be imposed (TeleGeography 2012b) which was subsequently specified as 50 % population coverage or fixed-wire coverage of 200 commercial and/or residential buildings (Telecom.paper 2012p). The auction was scheduled for 2013Q1.

Although Hutchison and PCCW own a joint licence and network in the 2.6 GHz spectrum band via Genius Brand, they intend to continue to launch independent services—PCCW first launched LTE in April 2012 as 'Navigator Everywhere' using as its fall-back its 42 Mbps DC-HSPA+ network combined with its 100 Mbps Wi-Fi service, with Hutchison following on in May (TeleGeography 2012j). For its part, SmarTone is restricted itself initially to the 1,800 MHz band with the intention of launching LTE in 2012H2. In June, CSL announced that it had signed a LTE roaming deal with South Korea's SK Telecom—followed in January 2013 by a similar agreement with Australia's Telstra.

In July, SmarTone launched LTE using its 1,800 MHz spectrum. It claimed that this had been a wise move because of the wide range of compatible devices becoming available capable of switching seamlessly to its HSPA+ 850/2,100 MHz network. Interestingly, it stated that unless customers were transferring files of at least 12 megabytes (a rare event using a smartphone), they would not notice any difference in transfer speeds during such switches. Rather, the virtue of LTE was its ability to extend broadband capacity, thereby preventing transfer speeds from slowing rapidly due to congestion. A roaming service with Singapore's MobileOne was initiated in September.

It is worth noting that at the time of its launch in August 2012, the CSL dual-band network branded as '4G LTE D-C' was capable of a downlink of 37 Mbps. Not only was this twice as fast as part-owner Telstra was providing in Australia at the time, but it was incredibly cheap for an 'all-you-can-eat' data plan (Mr Gadget 2012).

The 2.5/2.6 GHz auction eventually took place in March 2013. Predictably, the winners—paying so-called 'spectrum utilisation fees'—were SmarTone (two blocks at a cost of $82.5 million) and CSL, China Mobile and Genius Brand (one apiece for $39.9 million, $38.7 million and $37.4 million respectively) (Telecom. paper 2013e). Curiously, this was exactly the same amount as was raised in 2009 when a slab of 2.6 GHz spectrum was left unsold and furthermore, the 2.6 GHz spectrum was sold on both occasions for a very similar $/MHz/pop. On both occasions, this was extremely high for spectrum in this band, reflecting in good part, no doubt, the very compact land mass of Hong Kong. SmarTone acquired two blocks because it was the only one of the four winners not to hold existing spectrum in this band, but in the process it denied any spectrum to the fifth bidder, China Unicom.

In the aftermath of the auction, CSL's lifestyle brand one2free and mobile brand 1010 launched LTE extended spectrum meaning that, after the addition of 2,535–2,540 MHz paired with 2,655–2,660 MHz, a 20 MHz paired carrier was being employed in the 2.6 GHz band capable of running Category 4 devices—in this case a portable Wi-Fi router. A further 15 MHz paired in the 1,800 MHz band was subsequently added (Telecom.paper 2013h) with a projected upgrade to 20 MHz paired. In June, both PCCW and Hutchison added the matching 5 MHz paired acquired by Genius Brand to the pre-existing 15 MHz paired, and PCCW stated that its network was now capable of a maximum 150 Mbps downlink which could be accessed using the Huawei Ascend P2 Category 4 handset.

What will happen when 3G licences come up for renewal is less clear-cut. Standard practice elsewhere in the world is to renew these without controversy, but the regulator has stated its intention to renew only two-thirds of the spectrum while putting the rest up for auction, thereby creating an opportunity for a potential new entrant. Naturally, the incumbents are outraged, claiming that their networks would suffer serious degradation, and there is always the possibility that the matter will end up in the courts.

6.10 Indonesia

Indonesia is another country where the licensing process for 3G bandwidth has been ongoing and confused, and hence it is unclear whether any of the 3G frequencies will play a part in the provision of 4G. It is of interest that, as well as the assignment of 3G licences in the 1,900 MHz band, there was a licence in the 900 MHz band assigned to Indosat in September 2012. Furthermore, the Ministry issued a tender in the same month—postponed from a year earlier and pencilled in for March 2013— for so-called 'channel 11' and 'channel 12' frequencies in the 2,100 MHz band (TeleGeography 2012r). Two of the original twelve 1,900 MHz blocks initially left unsold, each of 5 MHz, were bought by Telkomsel and Axiata in March 2013 (TeleGeography 2013k).

The mobile operators listed in Table 1.2 are all involved in trials of LTE but the spectrum being used is unclear. The government has issued eight licences for WiMAX in the 2.3 GHz and 3.3 GHz bands and First Media, branded as Sitra, claimed to have launched 4G WiMAX in the 2.3 GHz band in July 2010. Mobile operator Bakrie Telecom acquired Reka Jasa Akses (REJA), which holds a national licence consisting of 12.5 MHz in the 3.3 GHz band, in August 2011. In October 2011, Indosat conducted a trial in the 1,800 MHz band using re-farmed spectrum, having previously conducted trials in the 2.6 GHz band during 2010.

In July 2012, the government announced that it would be monitoring developments overseas with a particular emphasis upon the spectrum bands being used in order to maximise opportunities for roaming and the availability of handsets. No launches could be expected before 2014, in good part because of the prior need to sort out ownership of suitable spectrum.

Recently, much thought has been given to the issue of an excessive number of operators. Currently, there are five GSM and three CDMA operators of any consequence but there simply isn't enough bandwidth to be divided up efficiently among so many operators—half as many would seem to be the optimal number according to the government. Hence, M&A activity is on the cards although specific proposals are thin on the ground with Axiata seemingly the likeliest to bid for a competitor (TeleGeography 2013k).

6.11 Japan

In August 2007, the regulator announced that it was ready to accept applications for licences in the 2.6 GHz band, with two licences to be awarded by the year-end. Priority was to be given to non-cellular operators, although incumbents would be able to take stakes of up to 33 % in consortia. KDDI bid with parent Kyocera, Intel and others acted jointly as UQ Communications, SoftBank bid with eMobile and DoCoMo bid with ACCA and others acting jointly as the ACCA Wireless Company. In December, the winners were announced as UQ Communications (802.16e) and Willcom (Super PHS), with each winning 30 MHz of unpaired spectrum, and the DoCoMo consortium was subsequently disbanded. For its part, UQ Communications is committed to WiMAX2, achieving 100 Mbps downstream during trials in 2011 and planning a service launch in 2013 (Telecom.paper 2011a).

Applications for new licences in Japan—in this case expressly for '3.9G'—were invited with a deadline of May 2009. These were awarded to incumbents DoCoMo, KDDI and SoftBank Mobile (each receiving 10 MHz in the 1.5 GHz band) as well as eMobile (10 MHz in the 1.7 GHz band) which committed to the launch of 3.9G LTE in 2010 and 4G LTE in 2012 but put back even the former date to 2012Q1. DoCoMo succeeded in launching Super 3G (branded as 'Xi'—pronounced 'Crossi') on 23 December 2010 using its 3G spectrum, providing 37.5 Mbps downstream and 12.5 Mbps upstream. As of the end of 2011Q2 it had acquired only 80,000 subscriptions due to the limited coverage of the network, but it had set in hand negotiations to lease the network to Japan Communications. For its part,

KDDI planned to use the 800 MHz band for nationwide coverage and the 1,500 MHz band in high-density urban areas at some future point, but it launched commercially in September 2012 using the 2.1 GHz band (Band 1) compatible with the iPhone 5 (Cellular-news 2012g). The projected coverage was 96 % of the population by April 2012.

According to one version, SoftBank had adopted a different strategy, opting in November 2011 to launch a Personal Handyphone System (PHS) network based on the Advanced eXtended Global Platform (AXGP) standard which is compliant with TD-LTE, and promising to cover 90 % of the population by the end of fiscal 2012 (TeleGeography 2011b). The 20 MHz of spectrum in the 2.5 GHz band was acquired when SoftBank bought Willcom. However, it was claimed elsewhere that it was TD-LTE which would be launched with backwards compatibility with PHS (Gabriel 2011). In the event, the launch in February 2012—via the licensee, the affiliated company Wireless City Planning—was announced as AXGP 'compatible with TD-LTE' (TeleGeography 2012d), with a launch of FDD using the 2.1 GHz band following in September (Telecom.paper 2012j) to coincide with the launch of the iPhone 5.

DoCoMo introduced a range of handsets that includes both variants of those available elsewhere such as the Samsung Galaxy S II, Sony Xperia GX and Xperia SX, and also some unique to itself such as the Fujitsu Arrows X LTE (F-05D) and Optimus LTE (L-01D). These helped it to achieve 1.14 million Xi customers by end-2011, rising to two million in March 2012, three million in June and four million in July. The latest models due for launch during 2013H1 can provide a maximum downlink of 112.5 Mbps and uplink of 37.5 Mbps and include the Aquos Phone EX (SH-04E), Medias X (N-04E), Medias W (N-05E), Eluga X (P-02E), Xperia Z (SO-02E), Arrows X (F-02E), Ascend D2 (HW-03E) and Optimus G Pro (L-04E). It is significant that it was only able to market the iPhone for the first time in September 2012 with the launch of the 4G-enabled iPhone 5 because Apple had finally agreed to make it work on the 2,100 MHz band which was the exclusive preserve of DoCoMo. KDDI has also introduced handsets specific to its network such as the Kyocera Digno S LTE.

In December 2011, Internet Initiative Japan (ILJ) announced that, starting in February 2012, it would be operating as a mobile virtual network enabler (MVNE), thereby speeding up the spread of LTE throughout Japan. It would operate over the DoCoMo network branded as 'Mobile/D' (Telecom.paper 2012b).

In January 2012, the government announced that it would be making available some spectrum in the 900 MHz band, with the licence winner paying to have the spectrum re-farmed. This was acquired by SoftBank.

For its part, eMobile (eAccess) announced in February 2012 that it would be launching in March a service using paired spectrum in the 1,700 MHz band (contained within LTE 1800), with a maximum downlink speed of 75 Mbps and a maximum uplink of 25 Mbps. All major cities would be covered by June. It added that it had successfully field-tested a (maximum) 300 Mbps service which it would be launching as LTE-Advanced in 2015. However, it noted that it was now the only operator not to own spectrum in the 800/900 MHz so-called 'platinum band' and hence that it would be applying for spectrum in the 700 MHz band (TeleGeography 2012f).

Japan has opted for the APT band plan for the 700 MHz band. In June, the Ministry duly awarded 700 MHz band spectrum to DoCoMo, KDDI and eMobile—SoftBank having already received 900 MHz band spectrum as noted above—although it was not expected to be utilised until early 2015 (TeleGeography 2012l).

In early October, SoftBank made an agreed takeover bid for eMobile (Katchi and Inagaki 2012) with completion taking place in January 2013. The implications of this for licences remain unresolved, but provided SoftBank reduces its stake to less than one-third of eMobile, as it is in the process of doing, both operators can retain their existing spectrum. Softbank also announced the launch of a LTE router capable of a theoretical 110 Mbps downlink. Also in October, SoftBank bid $20 billion for a 70 % stake in Sprint Nextel of the USA. Sprint Nextel immediately responded by raising its stake in TD-LTE satellite operator Clearwire (see Chap. 3) so that, in effect, SoftBank was acquiring control of both companies (Osawa 2012). In November, SoftBank formed a partnership with Thailand's TOT with a view to focussing on the development of TD-LTE in the 2.3 GHz band.

DoCoMo has long been a leader in the development of 3G femtocells—used for improved indoor coverage—and in December 2012 it launched a new plug-and-play device that would provide coverage simultaneously for both 3G and LTE (Cellular-news 2012k). It is also intent upon improving the LTE downlink—pencilling in a maximum of 112.5 Mbps in major cities by March 2014 and 187.5 Mbps nationwide during fiscal 2014. In July 2013, it was announced that a pre-launch of a 150 Mbps downlink would shortly be taking place in Kawasaki to be followed by the official launch in Tokyo and two other cities in October. In respect of the uplink, it experimented with a 400 MHz bandwidth in the 11 GHz band in Okinawa in December 2012, combining MIMO with eight transmitting antennas and 16 receiving antennas on the same frequencies. The aim was to achieve a 10 Gbps transfer, although spectrum bands above 5 GHz have never been favoured for mobile use given their short wavelengths and virtual inability to skirt around buildings.

More recently, it has established a lead in the testing of active antennas which feature energy-efficient operation and small form factors and are designed to overcome the difficulty that such antennas need to be made by the same vendor as the base stations to which they are linked. In future, it should be possible to set up the antennas independently of base stations (Cellular-news 2013b). In June 2013, it announced the widespread introduction of compact base stations, one-tenth the size of their predecessors, for use in low-traffic and difficult to access areas. The drawback is a halving of the maximum downlink and uplink.

In February 2013, DoCoMo signed up with NSN and Panasonic to roll out a LTE-A network capable of a 300 Mbps downlink. For its part, eAccess launched an Android Category 4 smartphone, the Huawei Stream X (GL07S), in March. However, there is evidence that operators are intent upon misleading customers about what their networks are capable of delivering in the real world. For example, KDDI advertised that virtually all customers with an iPhone 5 in major cities would be able to achieve a 75 Mbps downlink in March 2013. In practice, only 14 % of the population were able to achieve this on the given date and KDDI was given a regulatory reprimand for false advertising. It transpired that its claims related not to the iPhone 5 (2,100/700 MHz CDMA/LTE model) but to Android devices using the 800/1,500 MHz bands (TeleGeography 2013h). Despite this, KDDI announced that, commencing in June, it intended to upgrade its network to a maximum downlink of 100 Mbps.

6.12 Kiribati

Telecom Services Kiribati Ltd (TSKL) is to roll out a combined 3G/LTE network across the main urban areas of the 22 populated islands based on the Distributed Mobile Wireless Network technology supplied by Lemko and using the 700 MHz band. As the islands are only connected to each other and to the rest of the world via satellite, the main links will comprise LTE over satellite (Cellular-news 2013c).

6.13 Laos

Lao Telecommunications (LTC) is rolling out a LTE network while VimpelCom (Beeline) received a licence in May 2012 and is testing in Vientiane using the 1,800 MHz and 2.6 GHz bands.

6.14 Malaysia

The regulator appeared content to postpone the physical issue of licences beyond 2011 despite the fact that Packet One, for example, was ready to upgrade by overlaying its WiMAX network with TD-LTE while Asiaspace was considering the launch of TD-LTE in the 2.3 GHz band for which it had obtained a licence (with three others) in March 2007. The regulator's views on spectrum allocation were contained in its Spectrum Plan 2011 published in September 2011.

All the mobile incumbents either are conducting, or have completed, trials using paired spectrum. In October 2011, Maxis and U Mobile agreed to share their radio access network (RAN) outside the main urban areas once network roll-outs were authorised. Subsequently, the regulator announced that it was in favour of further partnerships for the delivery of 4G (Telecom.paper 2012f) and, in July 2012, Maxis and REDtone announced another arrangement to share spectrum and infrastructure over an initial 5-year period.

In December 2011, the regulator awarded seven 20 MHz licences in the 2.6 GHz band to mobile incumbents Axiata (Celcom), DiGi (49 % owned by Telenor), Maxis and U Mobile together with WiMAX operators Packet One Networks, REDtone and YTL Communications. In addition, 10 MHz was awarded to Asiaspace and 30 MHz to new entrant Puncak Semangat (via subsidiary Altel

Communications). The networks could be launched after 1 January 2013 (Telecom. paper 2011c). When the licences were officially handed over in December 2012, only eight were issued with Asiaspace omitted for undisclosed reasons (Telecom. paper 2012o) so Puncak Semangat ended up with 20 MHz paired, twice as much as its rivals. By June 2012, Axiata for one had announced that its entire network was now ready to provide 4G as soon as it received a licence, but it was Maxis that was the first to launch at the very beginning of 2013, providing up to a 75 Mbps downlink via dongles in selected parts of the Klang Valley (Telecom.paper 2013a). Axiata launched in the Klang Valley in April 2013 using dongles and predicted a nationwide roll-out by the end of 2015 (TeleGeography 2013f) with DiGi following suit in July (Telecom.paper 2013j).

Also in July, Axiata announced that it planned a joint roll-out with Puncak Semangat with each contributing 10 MHz paired—the other 10 MHz paired held by Puncak may be used in conjunction with another licensee. However, Axiata would act as the network operator with Puncak taking on the role of MVNO (TeleGeography 2013j).

6.15   Myanmar

In March 2013, the regulator made ready 10 MHz paired in the 1800 MHz band for LTE provision later in the year. However, it is first necessary to restructure the existing operations. The regulator intends to hive off its telecommunications operations into a new entity, Myanmar Telecom, which will not receive state subsidies. The other network, Yetanarpon Teleport, will become Yetanarpon Telecom. Two new unified licences will then be issued to companies that have at least a majority stake in an operator that already has at least four million subscribers and total revenues of at least $400 million. Because it is one of the last countries to open up its market, the opportunity to invest has attracted a huge number of interested parties (TeleGeography 2013d).

6.16   New Zealand

Licences covering between 20 MHz and 35 MHz unpaired in the 2.6 GHz band were issued in December 2007 to Vodafone plus four others. New spectrum in the form of 112 MHz in the 690–806 MHz digital dividend band was set to be auctioned in early 2013, with TNZ intent upon using this for LTE. UK-based Clear Mobitel also declared its intention to bid. However, in October 2012, a group of Maori interests made a Waitangi Tribunal claim just as they had done successfully—despite government opposition—at the time 3G licences were issued. In effect, the Maoris wanted to claim 4G spectrum as 'taonga' (cultural treasure) which the government vehemently opposed. Negotiations proved troublesome so any auction was set to be delayed (Telecom.paper 2012l)—the situation remains ongoing.

Telecom NZ (TNZ) commenced a two-stage trial, using the 2.6 GHz band in cities and the 700 MHz (APT band plan), 1,800 MHz and 2.6 GHz bands in rural areas, lasting from December 2012 to March 2013 with a view to a launch of LTE in Auckland in October 2013 followed by Wellington and Christchurch. In February 2013, the government announced that there would be an auction of blocks of spectrum in the 700 MHz band (APT band version). With the existing licences due to terminate at the year-end, the new licences would become operational on 1 January 2014 unless early access was negotiated in the interim. Maori interests would not be given a specific spectrum allocation but assisted in other ways (Cellular-news 2013d).

Also in February, Vodafone announced the launch in Auckland of a LTE network built by NSN using the 1,800 MHz band (Cellular-news 2013g), but is also running a rural trial in the 700 MHz band until the end of July. For its part, Woosh Wireless moved in March to consolidate its holdings of 2.3 GHz bandwidth via a purchase of '4G compatible' spectrum from Railway St Industries. As a result, Craig Wireless and its affiliates, trading jointly as the CWS Group, held spectrum in the 2.3 GHz, 2.5 GHz, 2.6 GHz and 3.5 GHz bands (TeleGeography 2013b). State-controlled Kordia and CWS Group are considering whether to switch their 2.3 GHz networks to LTE, and in April 2013 Craig Wireless acquired 35 MHz of this spectrum from Kordia in order to enhance the potential for such a switch.

6.17 Pakistan

The regulator set out to auction spectrum in the 1.9 GHz and 3.5 GHz bands in May 2012, divided respectively into regional blocks of 5 MHz and 21–23 MHz respectively. The auction would not be open to mobile operators, so it was unclear to what use the spectrum would be put (Telecom.paper 2012d), and it was also unclear how 4G would develop in the bands used elsewhere.

It has to be borne in mind that Pakistan only pencilled in an auction of 3G licences to take place in March 2012—almost the last major country to do so—and even then was unable to keep to that timetable. Three blocks of 9.8 MHz apiece will eventually be on offer.

6.18 Philippines

Smart Communications (the mobile subsidiary of PLDT) launched a modest commercial service in April 2011—referred to in certain quarters as a trial service—and set out to have its entire network ready to be switched to LTE—although HSPA+ would also be used as necessary—during 2012. A full commercial launch was duly claimed in August 2012 although some cities had been switched on at an earlier date (Cellular-news 2012e). Although it was reported that initially the 850 MHz and 2.6 GHz bands were used, with the 1,800 MHz band added in September making Smart the only tri-band 4G operator in the region, the reality is that the initial launch used the 2,100 MHz band with the 1,800 MHz and 850 MHz bands added in September (Villavicencio 2012). Smart signed a roaming agreement with Singapore's StarHub in October 2012 and launched its first handsets in December. During early 2013, it is experimenting with TD-LTE in conjunction with Huawei.

Globe Telecom, increasingly suffering from network congestion, brought forward its launch using the 1,800 MHz band to September and announced a new 'Tattoo Black' service in October. In October, it also announced the launch of data roaming arrangements with China Mobile (HK) and SK Telecom although the latter was also claimed to have occurred in April 2013. For its part, Piltel—which had transferred its existing spectrum to Smart Broadband post-merger—hopes to obtain a 3G licence and use the spectrum for LTE, unlike Bayan Telecommunications which intends to use existing 1,800 MHz spectrum while making use of Smart's 4G infrastructure in that band via a bilateral roaming agreement sanctioned by the regulator in early October (TeleGeography 2012s).

6.19 Singapore

In 2005, the regulator auctioned 10 blocks of 5 MHz in the 2.3 GHz band and 15 blocks of 6 MHz in the 2.6 GHz band specifically for WiMAX. Singapore Telecom (SingTel) and StarHub were capped at four blocks and other bidders at six blocks (deemed to be sufficient for a nationwide roll-out). SingTel bought four blocks in the 2.6 GHz band at a cost of S$2.05 million while StarHub merely bought two blocks in the 2.6 GHz band at a cost of S$1 million. In addition, MobileOne (M1) bought four blocks at a cost of S$2.1 million, inter-touch Holdings bought four blocks at a cost of S$1.02 million, Pacific Internet bought five blocks at a cost of S$2.27 million and Qala Singapore bought six blocks at a cost of S$1.40 million (Global View Partners 2009). One bidder was unsuccessful. However, no roll-outs took place. The issue was therefore whether to alter the licences of the six successful bidders to permit the use of LTE or to re-auction the spectrum.

All three mobile incumbents, SingTel, MobileOne (M1) and StarHub, commenced LTE testing early in 2010 using the 2.6 GHz band, but with the original WiMAX licences due to last until 2015 there was an urgent need to find a way for the incumbents to launch commercially at an earlier date. The regulator accordingly announced in January 2011 that the spectrum licensed in 2005 would be re-licensed in 2012 with a start date of 1 July 2015 when the 2005 licences expired, but any existing holder of a licence—which included all three incumbents—could proceed with a guarantee of being re-licensed in 2015. SingTel responded to this by declaring that it would launch by end-2011, while StarHub appeared likely to follow suit early in 2012 (Chua 2011). In the event, it was M1 that led the way with a commercial launch in June 2011 using both 1,800 MHz and 2.6 GHz bands, a strategy the others were likely to follow. Initially, however, it only served the business district.

In October 2010, three blocks of 5 MHz paired in the 3G (1,900/2,100 MHz) band were put up for auction, but as only the three incumbents applied, each was allocated an 11-year licence for the $20 million minimum (IDA 2010).

In April 2012, the regulator announced that a total of 320 MHz of spectrum in a variety of bands would be auctioned, of which 20 MHz paired in the 2.6 GHz band would be ring-fenced for a new entrant. The spectrum would largely consist of expiring licences—2.3 GHz (licensee QMax) and 2.6 GHz (licensees M1, StarHub, SingTel and PacketOne) on 1 July 2015 and 1,800 MHz (licensees M1, StarHub and SingTel) on 1 April 2017, with all expiring licences replaced with new ones that would expire in June 2,030. 70 MHz paired in the 1,800 MHz band and 60 MHz paired in the 2.6 GHz band would be set aside for FDD while 30 MHz in the 2.3 GHz band and 30 MHz in the 2.6 GHz band would be reserved for TD-LTE. No incumbent would be allowed to obtain more than a total of 90 MHz, and successful licensees would normally have 1 year from the date of issue of a licence to launch a network (TeleGeography 2012i).

In June, SingTel finally launched a service for users of smartphones, those available at the time being the HTC One XL, LG Optimus LTE and Samsung Galaxy S2 LTE. A real-world downlink of between 3.4 Mbps and 12 Mbps was forecast despite the theoretical maximum of 75 Mbps. It announced that it expected the nationwide roll-out to be completed in early 2013 (Telecom.paper 2012g). In July, M1 pre-empted this by stating that it expected to complete its nationwide roll-out before the end of 2012, at which point it would be introducing smartphones (TeleGeography 2012n). The launch duly took place in September (claiming to be the first in South East Asia) and also provided a roaming service with Hong Kong's SmarTone (Cellular-news 2012f). For its part, StarHub also launched in September albeit on a limited scale while promising to have half of the island covered by the year-end and nationwide coverage during 2013 (Cellular-news 2012i). In October, it signed a roaming agreement with Smart in the Philippines.

In an attempt to expand spectrum availability with a view to assisting in particular with access in remote areas, a Singapore White Spaces Pilot Group has been set up. The initial three pilot schemes took place in late 2012 using the 700 MHz band.

In January 2013, the regulator announced that the initial amount of spectrum to be made available for 4G would be increased to 150 MHz in the 1,800 MHz band together with 120 MHz in the 2.6 GHz band of which 40 MHz would be set aside for a new entrant with this being added to the main auction if no new entrant made a successful bid. All of the spectrum would be sold in blocks of 5 MHz paired at a reserve price of S$12.6 million for a 1,800 MHz block and S$7.9 million for a 2.6 GHz block, with the exception that a single new entrant bidder would get the 40 MHz reserved for S$40 million. Should there be two potential new entrants, the higher of the ensuing sealed bids would be given the licence while paying the amount offered by the other bidder. A new entrant could also compete in the main auction. All bids would be for spectrum in one band only. In the absence of a new entrant, an individual operator would be capped at 55 MHz paired, otherwise at 45 MHz paired. 2.6 GHz-only licensees would have to provide nationwide coverage at street level within 1 year and in underground stations and road tunnels within 3 years (Telecom.paper 2013b).

It is noteworthy that for the first time it was stipulated that licensees will be obliged to provide either LTE-Advanced or its WiMAX equivalent.

SingTel achieved its goal of nationwide coverage in April 2013, the second operator to achieve this goal after M1, in the process claiming a maximum downlink of 150 Mbps although this could not yet be accessed via handsets. It claimed to have signed up 300,000 LTE customers in the 10-month period since launch with the total for the whole of Singapore standing at roughly one million representing one in eight of total subscriptions.

The results of the auction were declared on 1 July. As there were no new entrants, all 270 MHz available were awarded to the three incumbents at the reserve prices. 150 MHz in the 1,800 MHz band was sold for US\$189 million and 120 MHz in the 2.6 GHz band was sold for \$95 million, raising \$289 million in total. Of this, \$81.9 million was paid by M1 for 40 MHz in both bands, \$107.1 million was paid by SingTel for 60 MHz in the 1,800 MHz band and 40 MHz in the 2.6 GHz band and \$94.4 million was paid by StarHub for 50 MHz in the 1,800 MHz band and 40 MHz in the 2.6 GHz band (Telecom.paper 2013i).

6.20 South Korea

South Korea is of particular interest because, in July 2012, it became the first country in the world to achieve 100 % wireless broadband penetration and it is claimed to have the world's fastest download speeds (Gabriel 2012b).

Unlike other countries, South Korea uses a proprietary version of WiMAX known as WiBro which it has upgraded via 'Wave 2' (alternatively 'Wave-II') to provide downstream speeds comparable to HSPA. With both technologies available and becoming ever faster, the need to move to LTE or LTE-Advanced was not perceived as urgent. KT Corp. (KT Telecom) was expected to be the first carrier in the world to offer nationwide 802.16e in June 2011 with LG Uplus following its example whereas, in contrast, SK Telecom (SKT) was undecided whether to switch its WiBro network to TD-LTE.

The operators had to re-apply for their 2G spectrum in April 2011, with the results of this likely to influence the introduction of LTE. In practice, with WiBro attracting only a small number of subscribers, SKT decided to move to LTE, launching initially in Seoul using the 800 MHz band in July 2011, with nationwide coverage to be achieved in 2013. Although an upgrade to LTE-Advanced is expected in 2013, SKT is also increasing its dependence upon Wi-Fi in congested areas (Snow 2011). For its part, LG Uplus announced a launch of LTE in 82 cities during 2011, initially in July, with nationwide coverage to be achieved by July 2012 (TeleGeography 2011a).

An auction for spectrum in the 800 MHz, 1,800 MHz and 2.1 GHz bands commenced at the end of June 2011, with SK Telecom and KT Corp. excluded from bidding for 2.1 GHz spectrum. In August, LG Uplus paid \$415 million for

20 MHz of 2.1 GHz band spectrum, KT Corp. paid \$243 million for 10 MHz in the 800 MHz band and SK Telecom paid \$925 million for 20 MHz in the 1,800 MHz band. KT Corp. announced that it would launch a full service in Seoul in November 2011, going nationwide by 2013, whereas both its rivals would be providing a data-only service (Wood 2011). It subsequently added that it would base its roll-out on its 'Warp' technology, developed in-house.

In September 2011, SKT claimed to have cancelled plans to introduce new LTE-enabled handsets because the regulator had refused it permission to raise prices compared to its equivalent 3G offerings. Despite this, it announced in November that the number of LTE subscribers was rising faster than expected—it reached 500,000 in mid-December and one million at end-January 2012—with the average user consuming roughly 50 % more data than the average 3G subscriber, and hence that it had brought forward to April 2012 its launch date for the first 84 cities. In May, it added that it had trialled a multi-carrier service which provided LTE simultaneously over the 800 MHz and 1,800 MHz bands, and that a commercial launch would take place in July (Cellular-news 2012d).

KT Corp. applied for permission to shut down its 2G service in order to re-farm the spectrum for 4G. Some of its subscribers took legal action to prevent this, but permission was granted in December and at the very beginning of 2012 the launch of LTE took place in Seoul using this spectrum. The switch-over was set to proceed steadily until nationwide provision of LTE became available in April 2012. It signed up its one-millionth customer in mid-June.

New WiBro licensees were expected to exert downward pressure on prices, but the technology now seems to have no real future given that, in December 2011, the regulator rejected for the second time the applications made by Internet Space Time and KMI.

LG Uplus announced that it would be introducing voice over LTE (VoLTE—see Chap. 1) in October 2012, much the same time as KT Corp.. Meanwhile, it achieved one million subscribers on its 800 MHz LTE network in mid-February, achieved national coverage of the population at the end of March (at a cost of \$1.1 billion) and announced that it would launch LTE using its 2.1 GHz spectrum in September.

In June, SKT announced that it had signed a LTE roaming deal with CSL in Hong Kong—another involving Globe in the Philippines was signed in April 2013. It went on to add that it planned to launch its VoLTE service in September in Seoul and six other large cities. It pointed out that since the existing 3G service, mVoIP, took the form of an Internet download whereas VoLTE would be embedded in handsets, the quality of a voice call would rise significantly. In practice, VoLTE was launched commercially in August branded as 'HD Voice', with potential users obliged to be signed up for LTE and in possession of a HD Voice-compatible device—initially this would be a Samsung Galaxy S III LTE handset. LG Uplus followed shortly thereafter in Seoul, using the LG Optimus LTE2. KT Corp. also plans to rely predominantly on Samsung Galaxy handsets for VoLTE.

Taken in conjunction with the plan to launch LTE-Advanced in 2013 (4G-Portal 2013), SKT referred to its overall programme as 'LTE 2.0'. Given that SKT is a leader in the field, with effective LTE nationwide coverage as of July 2012, it is worth itemising some technological aspects of this programme aside from VoLTE. Firstly, in July, it launched multi-carrier LTE as noted above—thereby in this case doubling its LTE spectrum to 40 MHz—and anticipated coverage of 24 major markets by early 2013 (Gabriel 2012a). Secondly, it planned the world's first commercial launch of a Hybrid Network Integrated Solution—which enables simultaneous use of LTE and Wi-Fi and hence supports a theoretical maximum downlink of up to 127 Mbps (LTE = 75 Mbps + Wi-Fi = 52 Mbps)—before the end of 2012. Thirdly, it planned in 2013H2 to launch Carrier Aggregation, a core LTE-Advanced technology which doubles transmission speeds—in this case by combining a theoretical maximum of 75 Mbps in the 800 MHz band with a theoretical maximum of 75 Mbps in the 1,800 MHz band. Although such technological leadership does not necessarily pay dividends, it is worth noting that, by July 2012, SKT's average LTE customer was transferring 60 % more data than its average 3G customer.

In December 2012, KT Corp. and LG Uplus announced an agreement to provide data sharing plans. In February 2013, KT Corp. launched unlimited data plans which could be construed as a reflection of intense competition in the market.

SK Telecom has been very successful in its attempt to be one of the first operators to establish Joyn (see above). A mere 50 days after its launch in December 2012, subscriber numbers exceeded one million.

The regulator intends to auction one 15 MHz block and one of 35 MHz in the 1,800 MHz band as well as two 40 MHz blocks in the 2.6 GHz band at some point beyond March 2013. It is considering whether to exclude SK Telekom and KT Corp. from taking part in the 1,800 MHz auction. KT Corp. reacted angrily, claiming that its two rivals, both subsidiaries of huge conglomerates, were seeking to exclude it in order to drive it from the mobile market. However, SK Telecom counter-attacked, arguing that if Korea Telecom were to be awarded additional 1,800 MHz spectrum—it was interested in the 15 MHz block—it would be able, unlike its rivals, to double its downlink speed cheaply and speedily (Telecom.paper 2013g).

In February 2013, SK Telecom felt obliged to respond to accusations that it had warned European operators about the 'curse' of 4G—for a discussion of the so-called 'winner's curse' see Curwen (2002: pp. 132–38)—claiming that the remark had in fact originated from a rival. Despite its protestations of innocence, it was observed that a senior manager had gone on record to the effect that whereas SK Telecom had increased its average revenue per unit (ARPU) by $13, this was barely sufficient to justify the investment in LTE. Interestingly, this remark took place in the context of an operator with 8.3 million subscribers at the time, equivalent to roughly 30 % of its overall subscriber base—in late April the number of subscribers crossed the 10 million mark equivalent to roughly 37 % of the total customer base. On a more positive note, SK Telecom also noted that once this had risen to 60 % by the year-end, the financial impact would be much more positive. Meanwhile, rival KT Corp. was bemoaning the huge increase in data transfers stimulated by the launch of its LTE network, and every operator was noting the negative consequence of subsidising new LTE devices.

At much the same time, operators were turning their thoughts towards the development of devices for LTE-A. In this context, Carrier Aggregation is a major concern and in February 2013 SK Telecom, for example, floated the need for devices that could combine 10 MHz of its 800 MHz spectrum with 10 MHz of its 1,800 MHz spectrum, thereby enabling a maximum downlink of 150 Mbps. The only smartphone capable of operating at this speed at the time was the Huawei Ascend P2. However, on 1 April, SK Telecom was able to launch some devices which were able to pick up (slower) signals in four bands simultaneously—800 MHz, 1,800 MHz, 2.1 GHz and 2.6 GHz (Telecoms.com 2013). LTE-A was officially introduced onto its network on 11 April, but the commercial launch was pencilled in for September.

In practice, the Samsung Galaxy S4 LTE-A was used in June 2013 when SK Telecom launched the world's first LTE-A service in central Seoul and 42 other cities providing a maximum downlink of 150 Mbps, twice that being provided previously—it had acquired 150,000 subscribers within 2 weeks of launch. The technology combined Carrier Aggregation with Coordinated Multi Point (CoMP) software, alternatively known as co-operative MIMO, and is to be upgraded with Enhanced Inter-Cell Interference Coordination (eICIC) (see Chap. 1) in 2014 (TeleGeography 2013i). It should be noted that this is in essence a process of adding technological advancements as part of a rolling programme. The switch to LTE from HSPA+ was a step-change in technology, but the upgrade to LTE-A does not follow the same pattern. There is no obligation on an operator to adopt a particular pathway (Gabriel 2013). LG Uplus followed suit in mid-July with one differentiating factor: its version of the Galaxy S4 LTE is coloured blue while that of SK Telecom is coloured red.

Also in June, wholesale carrier services provider BICS announced that it had used its IPX platform to enable inter-continental LTE roaming between South Korea and Switzerland. Canada and Hong Kong are to be added shortly.

In July, LG Uplus announced its own launch of LTE-A in Seoul and other unnamed major cities (TeleGeography 2013l).

6.21 Sri Lanka

Etisalat claimed that its HSPA+ network was 'ready for LTE' in May 2011—it launched DC-HSPA+ in September—at a time when rival incumbents Axiata (Dialog) and Mobitel were conducting trials of LTE. Despite an absence of announcements concerning new licences, the regulator claimed in September 2012 that all three would be in a position to launch LTE by the year-end. Such an early launch was unlikely to find favour with many, if any, of the five incumbents, and in any event the plan was to auction the licences at the end of 2012 with commercial launches pencilled in for 2013 (TeleGeography 2012t). However, Huawei claimed that Axiata had launched TD-LTE using its equipment in Colombo on the final day of 2012 via subsidiary Dialog Broadband Networks, but did not specify the spectrum band (Huawei 2013). It later turned out to be the case that this

was a TD-LTE network consisting of 15 MHz of fixed-wireless provision in the
2.3 GHz band and hence not qualifying as a mobile launch. An additional block of
2.3 GHz spectrum was bought in May 2013. Interestingly, Mobitel also announced
that it had launched on the same day in December, again without specifying the
spectrum band although this appears to have been 1,800 MHz, with both claiming
to have been the first to launch.

At the end of March 2013, the regulator announced that Axiata had won an
auction for a 10-year licence providing 10 MHz in the 1,800 MHz band with a bid
of $25.3 million. In early April, Axiata launched LTE in Colombo using this
spectrum (TeleGeography 2013e).

6.22 Taiwan

According to reports, the regulator was initially in no rush to issue licences with the
700 MHz band being favoured. However, in March 2011, it was alleged that First
International Telecom (Fitel) was about to commence TD-LTE trials in conjunction
with China Mobile. A provider of PHS in the 1,900 MHz band, Fitel was granted a
WiMAX licence in 2007 which the regulator insists is tied explicitly to WiMAX
even though most of the other WiMAX licensees (FarEasTone, Global On Corporation, Tatung Telecom and Vmax Telecom) were given permission to switch to
LTE in June 2010—there are very few WiMAX subscribers (only 133,000 at end-
2011) but Taiwanese vendors make and export WiMAX equipment so the government is reluctant to promote LTE. Like the WiMAX operators, Vastar Cable TV
Systems also received 30 MHz of unpaired spectrum. In April 2011, mobile
incumbent Chunghwa Telecom announced that it had completed a 6-month trial
using both the 700 MHz and 2.6 GHz bands.

In November 2011, possibly in response to progress in South Korea (see above),
the regulator announced that it now intended to auction six licences for LTE during
2013, 2 years earlier than its previous target of July 2015. This would involve
spectrum in the 700 MHz band currently used for military training exercises,
between 10 and 15 unused blocks in the 2.6 GHz band and re-farmed 2G spectrum
(TeleGeography 2011d). However, it back-tracked in December on the grounds that
it was proving difficult to reorganise the relevant spectrum bands, stating that no
decisions had yet been made about technology use, that an auction would take place
in 2015 and that launches would take place in 2017 (Telecom.paper 2011b). In
February 2012, the plan had once again evolved to comprise the issue of at least
three 15-year licences within 3 years, and a government spokesman subsequently
stated that any outcome would need to be fair in relation to the implications for
WiMAX licensees.

In March, it was back to Plan A, with licence issue in July 2013 and launches by
July 2015—or possibly at some other time depending on which organisation the
spokesman in question was representing. In mid-June, the Ministry of Transport
and Communications set out a proposal to issue eight licences, to be submitted to
the Cabinet in July. This proposal recommended an auction of spectrum in the
800 MHz, 900 MHz and 1,800 MHz bands on a technology-neutral basis, and

Table 6.1 Planned auction. Taiwan. 2013

Bandwidth	700 MHz Number	700 MHz Min. Price	900 MHz Number	900 MHz Min. Price	1,800 MHz Number	1,800 MHz Min. Price
10 paired	3	$153 mn	2	$70 mn	3	$47 mn
10 paired			1	$53 mn		
15 paired	1	$230 mn			1	$73 mn
15 paired					1	$100 mn

would effectively force existing licensees such as Chunghwa Telecom and FarEasTone, whose 2G licences were approaching renewal, to enter bids in order to guarantee continuance of their 2G/3G services.

In October, the regulator specified that new licences would outlaw the use of unlimited data plans on the grounds that 15 % of existing users were consuming 65 % of the available bandwidth. In addition, operators with ideas for Taiwan's creative industries would be given a discount on the licence fee.

In January 2013, the regulator announced that it intended to auction 270 MHz of spectrum in the 700 MHz, 900 MHz and 1,800 MHz bands. The maximum amount of spectrum to be bought by any operator would be determined by the number of bidders—if more than five bidders participated than each would be capped at seven blocks of 5 MHz paired, whereas fewer than five bidders would result in the cap being raised to eight or nine blocks (Telecom.paper 2013c). In May, it was announced that spectrum would be awarded in blocks of 10 MHz paired or 15 MHz paired, with the reserve prices fixed as follows: $153 million per 20 MHz block and $230 million per 30 MHz block in the 700 MHz band; $53 million per 20 MHz block in the 900 MHz band; $47 million per 20 MHz block and $73 million per 30 MHz block in the 1,800 MHz band (Telecom.paper 2013f).

In February, Fitel was authorised to upgrade from PHS to XGP (following in the footsteps of, and co-operating with, Softbank/Willcom in Japan) with a view to a launch in 2014 (Telecom.paper 2013d). However, issues were raised elsewhere concerning the potential launch date using spectrum previously in use for 2G, given that the relevant licences were not due to expire until 2017.

In June, seven potential licensees were revealed as incumbents Chunghwa Telecom, Taiwan Mobile, FarEasTone and Asia Pacific Telecom together with Foxconn Group subsidiary Ambit Microsystems, Ting Hsin International Group and Shinkong Group. Some refinements were made to the previously announced minimum prices to allow for differing dates of availability, as shown in Table 6.1.

6.23 Thailand

The situation in Thailand is too complicated to expound fully here, in good part because of the existence of state-owned CAT and TOT and their revenue-sharing build-transfer-operate (BTO) arrangements with the private operators. However, the following can be taken into account.

In June 2011, DTAC offered to return an unused 25 MHz block in the 1,800 MHz band—it owned 50 MHz in the 1,800 MHz band for voice and 10 MHz in the 850 MHz band for data—in return for a 4G licence. 12.5 MHz in the 1,800 MHz band were due to be returned by operators when their licences came up for renewal—True Move and AIS in September 2013, DTAC in 2018. Adding 25 MHz to 12.5 MHz paired would allow a 4G auction with licences rather than BTO arrangements, possibly in 2014Q3 after 3G licences had been assigned.

Meanwhile, in December 2011, responding to the regulator's announcement that it would allow for a 90-day period of 4G testing over TOT's spectrum, AIS arranged to trial LTE using spectrum in the 1,800 MHz band owned by CAT and (acting as Digital Phone) in the 2.3 GHz band owned by TOT. True Move also indicated that it might follow suit (TeleGeography 2012a).

In March 2012, TOT announced that, subject to regulatory permission, it would be returning 34 MHz of non-contiguous 2.3 GHz spectrum to the regulator while retaining the remaining contiguous 64 MHz (not involved in BTO concessions) for its own 4G network, which it would then operate as a wholesaler over a 15-year period (TeleGeography 2012e). This was agreed in the regulator's 'master plan' published shortly thereafter (TeleGeography 2012g) and TOT has subsequently announced that it intends to develop the spectrum with the help of Japan's SoftBank.

In May, the regulator authorised DTAC and True Move to proceed with 4G trials in the 1,800 MHz band. The report added, confusingly, that AIS had already received permission to conduct trials in the 900 MHz band.

In September, TOT announced that it would be conducting a trial before the year-end using spectrum in the 2,300 MHz band. Two private operators would also participate, with both placing equipment at 200 sites and using the TOT network to supply services. TOT put in a request to the regulator for permission to utilise a 30 MHz block of its remaining 64 MHz of spectrum in this band. In a further move, the regulator rescinded the previous decision that AIS and True Move return their 1,800 MHz licences in September 2013, although the length of any licence extension was not specified.

It is now being mooted that the regulator will extend the 2G licences of AIS and True Move until 2018 which would allow for a 62.5 MHz block in the 1,800 MHz band to be auctioned off in one go, possibly in conjunction with the 900 MHz spectrum due to be returned by DTAC if extended from September 2015 (Cellularnews 2012h).

It should be borne in mind that at the time when the sale of 4G licences was being addressed, the same process was taking place in respect of 2.1 GHz spectrum intended to be used for 3G although it could in principle be used for 4G as it was technology-neutral. The unusual structure of the mobile market had caused the 3G auction to be delayed on many occasions, with a date finally pencilled in for October 2012. Nine blocks, each of 5 MHz, were on offer via 15-year Type 3 (infrastructure-based as against BTO) licences with any successful bidder capped at

15 MHz. On the face of it, 17 companies from 11 groups appeared to be preparing bids, but as TOT subsidiary ACT Mobile and Cable Thai Holding were intent upon the role of observer while three others were subsidiaries of AIS (which bid in the end via Advanced Wireless Network), three were subsidiaries of True Group (which bid in the end via Real Future) and two were subsidiaries of DTAC (which bid in the end via DTAC Network), it was suspected that, in reality, there would only be three bidders (Telecom.paper 2012h). In the event, Tantawan Telecommunications also applied but was rejected. With only three bidders, $1.36 billion was raised compared to a floor set at $1.32 billion. Only AIS offered slightly above the floor price because it wanted to acquire the 1,950–1,965 MHz and 2,140–2,155 MHz bands that were contiguous to BTO partner TOT's 1,965–1,980 MHz and 2,155–2,170 MHz bands.

Licences were meant to be issued 1 week after payment was made, but no later than January 2013 and launches were expected fairly soon thereafter. However, after accusations surfaced claiming that proper procedures had not been followed and that the three licensees had colluded, the regulator opened an investigation which, if it dragged on beyond January 2013, would cause the licences to be annulled (Cellular-news 2012j). In December, the regulator ruled that the licences would be issued with the proviso that voice and data tariffs would have to be 15 % lower than the average on 7 December. A group of senators promptly appealed the decision (TeleGeography 2012w) and in June 2013—speed is seemingly not of the essence—the National Anti-Corruption Commission ruled that there had been irregularities in the auction design and stated that proceedings might be taken against various parties involved.

Meanwhile, in November 2012, DTAC announced that it would launch 3G by June 2013 provided it received its licence in 2012; build out its networks in both the 1,800 MHz and 850 MHz bands to provide 3G coverage for half of the population during 2014; and launch LTE using the 1,800 MHz band as soon as it received a licence (Telecom.paper 2012m). For its part, AIS, which already provided a nationwide 3G service using the 900 MHz band, launched 3G in May using the 2.1 GHz band. True Move, which already provided a nationwide 3G service using the 850 MHz band, achieved a limited launch of 3G and LTE combined in May 2013 in parts of Bangkok using 2.1 GHz spectrum, but there is no intention to launch nationwide even though the target for LTE subscription for the end of 2013 is six million. As for TOT, it was claimed that it would be awarding a trial LTE licence in the 2.1 GHz band to equipment and services group Samart in February 2013. The proposal was for Samart to install LTE equipment at 100 base stations as part of TOT's 3G trial. However, no other companies are interested in operating as a MVNO over the TOT network.

Rather alarmingly, especially given that operators' 3G licences only specified a 153 kbps uplink and 345 kbps downlink, complaints began pouring in that the new 3G networks were often slower than existing 2G networks. The comparison with South Korea could not be more stark.

6.24   Vietnam

In 2010, five operators were given permission to conduct trials—Antares (in association with VNPT), CMC Telecom, FTP Telecom, Viettel and VTC. The trials were expected to last for 1 year followed by an auction at the end of that period. However, in February 2012, the government decided to delay the sale of licences until 2018 on the grounds that 3G had been slow to take off (Telecom. paper 2012c). In September 2012, the regulator reiterated that sales of 3G subscriptions had remained sluggish and announced that since there was no immediate need for 4G, it would not be licensed until 2015 (Telecom.paper 2012i).

References

4G-Portal. (2013). *SK Telecom and Ericsson on the way towards LTE-A.* Accessed January 31, 2013, from http://4g-portal.com

Aetha. (2011). *Case studies for the award of the 700MHz/800MHz band: Australia.* Accessed November 20, 2011, from http://www.gsma.com/spectrum

Cellular-news. (2010). *Hong Kong plans radio spectrum auction in the 850 MHz, 900 MHz and 2 GHz bands.* Accessed December 11, 2010, from http://www.cellular-news.com

Cellular-news. (2011). *Optus plans LTE network launch in early 2012.* Accessed September 16, 2011, from http://www.cellular-news.com

Cellular-news. (2012a). *Australia outlines plans for digital dividend spectrum auction.* Accessed April 11, 2012, from http://www.cellular-news.com

Cellular-news. (2012b). *ZTE passes second phase of China TD-LTE test.* Accessed May 30, 2012, from http://www.cellular-news.com

Cellular-news. (2012c). *Vodafone Australia to deploy LTE services next year.* Accessed June 26, 2012, from http://www.cellular-news.com

Cellular-news. (2012d). *SK Telecom commercially launches multi-carrier LTE services.* Accessed July 3, 2012, from http://www.cellular-news.com

Cellular-news. (2012e). *Philippines Smart to launch LTE services this week.* Accessed August 20, 2012, from http://www.cellular-news.com

Cellular-news. (2012f). *MobileOne to launch nationwide LTE network next week.* Accessed September 9, 2012, from http://www.cellular-news.com

Cellular-news. (2012g). *Two Japanese LTE network launches due this friday.* Accessed September 17, 2012, from http://www.cellular-news.com

Cellular-news. (2012h). *Thailand's regulator mulls extending 2G spectrum licences.* Accessed October 2, 2012, from http://www.cellular-news.com

Cellular-news. (2012i). *Singapore's StarHub plans LTE launch next week.* Accessed October 2, 2012, from http://www.cellular-news.com

Cellular-news. (2012j). *Thai regulator unsure when it can formally award 3G licences.* Accessed November 5, 2012, from http://www.cellular-news.com

Cellular-news. (2012k). *DoCoMo develops first small-cell base station for 3G and LTE.* Accessed November 19, 2012, from http://www.cellular-news.com

Cellular-news. (2013a). *Bangladesh changes 3G licence tender to permit more foreign investors.* Accessed January 31, 2013, from http://www.cellular-news.com

Cellular-news. (2013b). *Japan's NTT DoCoMo tests active antenna on live LTE network.* Accessed February 21, 2013, from http://www.cellular-news.com

Cellular-news. (2013c). *Pacific island nation of Kiribati to get 3G/LTE network.* Accessed February 21, 2013, from http://www.cellular-news.com

Cellular-news. (2013d). *New Zealand plans 700 MHz license auction later this year*. Accessed February 21, 2013, from http://www.cellular-news.com

Cellular-news. (2013e). *Afghanistan regulator denies issuing a 4G license to Etisalat*. Accessed February 21, 2013, from http://www.cellular-news.com

Cellular-news. (2013f). *Ericsson to deploy LTE network for Bhutan Telecom*. Accessed February 28, 2013, from http://www.cellular-news.com

Cellular-news. (2013g). *Vodafone New Zealand launches LTE service in Auckland*. Accessed February 28, 2013, from http://www.cellular-news.com

Chua M. (2011). *Singapore 4G auction slated for 2012*. Accessed May 30, 2011, from http://www.telecomseurope.net

Curwen, P. (2002). *The future of Mobile Communications: Awaiting the third generation*. Basingstoke: Palgrave Macmillan.

Gabriel C. (2011). *Softbank and China Mobile tout TD-LTE progress*. Accessed November 16, 2011, from http://www.rethink-wireless.com

Gabriel C. (2012a). *SKT pushes multicarrier LTE forward*. Accessed June 1, 2012, from http://www.rethink-wireless.com

Gabriel C. (2012b). *South Korea tops 100% wireless broadband*. Accessed July 24, 2012, from http://www.rethink-wireless.com

Gabriel C. (2013). *SKT and Samsung team on 'LTE-A' launch*. Accessed June 26, 2013, from http://www.rethink-wireless.com

Global View Partners. (2009). *The 2.6 GHz spectrum band, report*. Prepared for the GSM Association, December.

Globaltelecomsbusiness. (2011). *Hong Kong awards two spectrum licences*. Accessed April 19, 2011, from http://www.globaltelecomsbusiness.com

Huawei. (2013). *Huawei LTE TDD technology powers Sri Lanka's first LTE network launched by Dialog*. Accessed January 22, 2013, from http://www.huawei.com

IDA. (2010). *IDA grants 3G spectrum rights (2010)*. Accessed October 29, 2010, from http://www.ida.com

Katchi H, Inagaki K. (2012). *Softbank to acquire eAccess via $2.3bn stock swap*. Accessed October 1, 2012, from http://www.cellular-news.com

Mr Gadget. (2012). *Review of world's first LTE D-C network by CSL 1010*. Accessed August 24, 2012, from http://www.mrgadget.com.au

Osawa J. (2012). *Softbank deal for Sprint turns on spectrum*. Accessed October 21, 2012, from http://online.wsj.com

Porter C. (2013). *Vodafone 4G network launches across Australia today*. Accessed July 10, 2013, from http://www.news.com.au

Snow C. (2011). *SK Telecom announces specifics for the first South Korean LTE network*. Accessed May 30, 2011, from http://www.telecomengine.com

Tawakilagi. (2012). *Unwired Fiji launched in the West*. Accessed November 21, 2011, from http://www.tawakilagi.com

Taylor J. (2013). *Telstra, Optus, TPG win 4G spectrum for AU$2 billion*. Accessed May 7, 2013, from http://www.zdnet.com

Telecom.paper. (2011a). *UQ achieves 100 Mbps transmission in Wimax 2 trial*. Accessed July 7, 2011, from http://www.telecompaper.com

Telecom.paper. (2011b). *NCC slammed for slow 4G development*. Accessed December 6, 2011, from http://www.telecompaper.com

Telecom.paper. (2011c). *Nine players to receive 2.6 GHz spectrum – report*. Accessed December 12, 2011, from http://www.telecompaper.com

Telecom.paper. (2012a). *Hong Kong gets new mobile entrant in LTE auction*. Accessed February 7, 2012, from http://www.telecompaper.com

Telecom.paper. (2012b). *IIJ to sell NTT Docomo LTE service from 27 February*. Accessed February 16, 2012, from http://www.telecompaper.com

Telecom.paper. (2012c). *Vietnam puts 4G licensing on hold.* Accessed February 21, 2012, from http://www.telecompaper.com

Telecom.paper. (2012d). *PTA issues details on 1.9, 3.5GHz bands auction.* Accessed March 2, 2012, from http://www.telecompaper.com

Telecom.paper. (2012e). *Regulator completes 3G licensing draft.* Accessed March 26, 2012, from http://www.telecompaper.com

Telecom.paper. (2012f). *Regulator sees partnerships for 4G rollouts.* Accessed April 18, 2012, from http://www.telecompaper.com

Telecom.paper. (2012g). *SingTel launches LTE services, fresh data bundles.* Accessed June 5, 2012, from http://www.telecompaper.com

Telecom.paper. (2012h). *10 firms show interest in 3G auction.* Accessed September 13, 2012, from http://www.telecompaper.com

Telecom.paper. (2012i). *Vietnam to delay 4G licensing until 2015.* Accessed September 25, 2012, from http://www.telecompaper.com

Telecom.paper. (2012j). *Softbank unveils LTE plans, voice options.* Accessed October 25, 2012, from http://www.telecompaper.com

Telecom.paper. (2012k). *DoCoMo Pacific to launch LTE in Guam.* Accessed October 4, 2012, from http://www.telecompaper.com

Telecom.paper. (2012l). *Maori claim could delay 4G spectrum allocation.* Accessed October 4, 2012, from http://www.telecompaper.com

Telecom.paper. (2012m). *DTAC tests 4G mobile service.* Accessed November 9, 2012, from http://www.telecompaper.com

Telecom.paper. (2012n). *Comms minister intervenes in 4G spectrum auction.* Accessed November 21, 2012, from http://www.telecompaper.com

Telecom.paper. (2012o). *Maxis ready to roll out LTE 2600.* Accessed December 17, 2012, from http://www.telecompaper.com

Telecom.paper. (2012p). *Hong Kong invites bid for radio spectrum for mobile services.* Accessed December 21, 2012, from http://www.telecompaper.com

Telecom.paper. (2013a). *Maxis launches LTE services in Klang Valley.* Accessed January 2, 2013, from http://www.telecompaper.com

Telecom.paper. (2013b). *Singapore plans 4G spectrum auction in mid-2013.* Accessed January 21, 2013, from http://www.telecompaper.com

Telecom.paper. (2013c). *NCC to auction more spectrum this year.* Accessed January 25, 2013, from http://www.telecompaper.com

Telecom.paper. (2013d). *Fitel gets green light to upgrade network to XGP.* Accessed February 11, 2013, from http://www.telecompaper.com

Telecom.paper. (2013e). *Hong Kong awards 2.5/2.6 GHz spectrum to 4 operators.* Accessed March 20, 2013, from http://www.telecompaper.com

Telecom.paper. (2013f). *NCC sets reserve price for 4G spectrum.* Accessed May 17, 2013, from http://www.telecompaper.com

Telecom.paper. (2013g). *SK Telecom, LG Uplus call for fair spectrum issue.* Accessed May 21, 2013, from http://www.telecompaper.com

Telecom.paper. (2013h). *1010, one2free launch LTE extended spectrum.* Accessed May 24, 2013, from http://www.telecompaper.com

Telecom.paper. (2013i). *M1, SingTel, StarHub secure 4G spectrum.* Accessed July 1, 2013, from http://www.telecompaper.com

Telecom.paper. (2013j). *DiGi launches LTE in Klang Valley.* Accessed July 8, 2013, from http://www.telecompaper.com

Telecoms.com. (2013). *SKT launch quad band devices on 1 April.* Accessed March 28, 2013, from http://www.telecoms.com

TeleGeography. (2011a). *SKT and LG Uplus inaugurate commercial LTE services.* Accessed July 4, 2011, from http://www.telegeography.com

TeleGeography. (2011b). *Softbank to launch 100Mbps PHS mobile data service.* Accessed October 4, 2011, from http://www.telegeography.com

TeleGeography. (2011c). *VelaTel and NGSN to launch 4G in China.* Accessed October 26, 2011, from http://www.telegeography.com

TeleGeography. (2011d). *NCC will put six LTE licences up for grabs in 2013.* Accessed November 8, 2011, from http://www.telegeography.com

TeleGeography. (2012a). *Operators invited to test LTE using TOT's spectrum.* Accessed January 9, 2012, from http://www.telegeography.com

TeleGeography. (2012b). *OFTA consults on spare LTE spectrum auction.* Accessed January 9, 2012, from http://www.telegeography.com

TeleGeography. (2012c). *Comms minister directs ACMA to limit digital dividend spectrum bidding.* Accessed February 9, 2012, from http://www.telegeography.com

TeleGeography. (2012d). *Softbank sets Friday as 4G launch date.* Accessed February 21, 2012, from http://www.telegeography.com

TeleGeography. (2012e). *TOT proposes 2.3GHz LTE, tower sharing strategies.* Accessed March 20, 2012, from http://www.telegeography.com

TeleGeography. (2012f). *Second LTE mobile network in Japan launched by eMobile.* Accessed March 21, 2012, from http://www.telegeography.com

TeleGeography. (2012g). *NBTC approves master plan: Spectrum conditions favour TOT.* Accessed March 22, 2012, from http://www.telegeography.com

TeleGeography. (2012h). *China Mobile Hong Kong LTE launch scheduled for Q4.* Accessed March 22, 2012, from http://www.telegeography.com

TeleGeography. (2012i). *Singapore's IDA to auction 4G spectrum from 2013.* Accessed April 16, 2012, from http://www.telegeography.com

TeleGeography. (2012j). *SAR Wars 4G: Cellcos do battle over LTE.* Accessed May 5, 2012, from http://www.telegeography.com

TeleGeography. (2012k). *Optus, VHA to share mobile base stations.* Accessed May 5, 2012, from http://www.telegeography.com

TeleGeography. (2012l). *DoCoMo reveals plans to 'expand and improve' Xi LTE network.* Accessed June 29, 2012, from http://www.telegeography.com

TeleGeography. (2012m). *Telstra examining options for LTE Advanced to keep competitive edge.* Accessed July 3, 2012, from http://www.telegeography.com

TeleGeography. (2012n). *M1 targets nationwide 4G coverage for dongles/smartphones in Q3, paper says.* Accessed July 16, 2012, from http://www.telegeography.com

TeleGeography. (2012o). *China Mobile HK awards ZTE, Ericsson TD-LTE contracts.* Accessed July 19, 2012, from http://www.telegeography.com

TeleGeography. (2012p). *MoPT announces draft proposals for 3G/4G spectrum auction.* Accessed July 24, 2012, from http://www.telegeography.com

TeleGeography. (2012q). *IT&E launches LTE: GTA boosts network coverage.* Accessed July 30, 2012, from http://www.telegeography.com

TeleGeography. (2012r). *MoCI sets September date for 3G channel sale; sets fee at USD20m.* Accessed September 11, 2012, from http://www.telegeography.com

TeleGeography. (2012s). *Bayan turns its back on NTC's 3G licence auction.* Accessed October 29, 2012, from http://www.telegeography.com

TeleGeography. (2012t). *Sri Lanka to auction 4G licences this year.* Accessed November 16, 2012, from http://www.telegeography.com

TeleGeography. (2012u). *ATTI launches 1800MHz consultation.* Accessed December 3, 2012, from http://www.telegeography.com

TeleGeography. (2012v). *18 December TDD/FDD LTE launch date for China Mobile Hong Kong.* Accessed December 5, 2012, from http://www.telegeography.com

TeleGeography. (2012w). *NBTC issues 3G licences; group of senators appeal.* Accessed December 10, 2012, from http://www.telegeography.com

TeleGeography. (2013a). *Vodafone Hutchison Australia kicks off LTE trials in Sydney*. Accessed February 19, 2013, from http://www.telegeography.com

TeleGeography. (2013b). *All aboard the spectrum gravy train as Woosh completes 2.3GHz deal*. Accessed March 18, 2013, from http://www.telegeography.com

TeleGeography. (2013c). *iConnect launches LTE in Guam*. Accessed March 21, 2013, from http://www.telegeography.com

TeleGeography. (2013d). *1800MHz band standardised for FD-LTE in Myanmar; trial expected in 2013*. Accessed March 21, 2013, from http://www.telegeography.com

TeleGeography. (2013e). *Dialog Axiata launches mobile FDD-LTE following licence win*. Accessed April 2, 2013, from http://www.telegeography.com

TeleGeography. (2013f). *Celcom introduces commercial LTE services in Klang Valley*. Accessed April 24, 2013, from http://www.telegeography.com

TeleGeography. (2013g). *800MHz spectrum frozen by apex court; Bangla-Russian venture applies for 2.3GHz mobile WiMAX concession*. Accessed May 15, 2013, from http://www.telegeography.com

TeleGeography. (2013h). *KDDI upgrades to 100Mbps; 800MHz, 2100 MHz LTE reaches 97%, 71% of population*. Accessed June 20, 2013, from http://www.telegeography.com

TeleGeography. (2013i). *SK Telecom inaugurates commercial LTE-A network*. Accessed June 26, 2013, from http://www.telegeography.com

TeleGeography. (2013j). *LTE joint rollout inked by Celcom, Puncak Semangat*. Accessed July 5, 2013, from http://www.telegeography.com

TeleGeography. (2013k). *MoCI official says telecoms giants see mergers as way to uncork the spectrum bottleneck*. Accessed July 11, 2013, from http://www.telegeography.com

TeleGeography. (2013l). *LG Uplus introduces LTE-A commercial services*. Accessed July 18, 2013, from http://www.telegeography.com

TeleGeography. (2013m). *Fiji kicks off 4G spectrum auction; Vodafone fills 30MHz quota*. Accessed July 22, 2013, from http://www.telegeography.com

Villavicencio P. (2012). *Smart: We will support all LTE devices running on our bands*. Accessed September 15, 2012, from http://www.interaksyon.com

Wood N. (2011). *KT Corp eyes November LTE launch*. Accessed February 16, 2012, from http://www.totaltele.com

Africa, Middle East and the Americas

7.1 Africa

7.1.1 Algeria

Algeria is one of the most backward countries in relation to high-speed data. It is still grappling with 3G, having postponed the issue of licences several times since the original planned date of October 2011. In April 2013, it was announced by the government that it was waiting to conclude the acquisition of a majority stake in Djezzy before proceeding since a company undergoing a change in shareholder structure is unable bid for a licence and this would place the now state-owned Djezzy at a disadvantage (Telecom.paper 2013c). Once this had been completed, Djezzy, Nedjma and Algérie Télécom Mobile (Mobilis), which is wholly state-owned, were advised to ready themselves for the issue of licences.

In July 2013, the government announced that the auction would begin on 1 August and that it was hoped that services would commence on 1 December irrespective of what was happening in relation to Djezzy.

7.1.2 Botswana

A trial by Mascom (majority-owned by MTN) commenced in June 2012 in Gaborone using its technology-neutral 2007 licence.

7.1.3 Ghana

In March 2013, the government announced that it would not make a formal allocation of 4G licences, but existing licensees were welcome to re-farm their spectrum. So far, no incumbent has taken up this opportunity, and it is possible that the first LTE networks to launch will be fixed-wireless and provided by new entrants. In July 2013, Surfline Communications (about which information is thin on the ground)

P. Curwen and J. Whalley, *Fourth Generation Mobile Communication*,
Management for Professionals, DOI 10.1007/978-3-319-02210-9_7,
© Springer International Publishing Switzerland 2013

selected Alcatel-Lucent to roll out its LTE network, initially launching in Accra in early 2014.

7.1.4 Kenya

Although Safaricom initially conducted separate testing using its 3G spectrum, it was decided that operators would not be awarded individual licences. Rather, the part-state-owned company would build a shared network and act as a wholesaler. However, in September 2011, the government announced that access to the network would be limited to companies holding a Network Facilities Provider Tier 1 licence, companies able to provide nationwide services within 1 year and companies with 20 % local ownership—thereby ruling out Bharti Airtel and Essar Telecom Kenya (yu). These requirements effectively limited participation to Safaricom and Orange (Telkom Kenya) (TeleGeography 2011a).

No intentions to participate were tabled by the 13 September deadline so it was extended to 27 September, by which point 17 parties had expressed an interest in response to a revised rule book. Amongst these were numbered eight international vendors, although they had yet to form consortia with local companies which the rules required to take place prior to February 2012, with a 20 % stake to be transferred to the local partner within 3 years. By late November, two vendors and four local ISPs had dropped out, although one additional vendor had expressed an interest.

In March 2012, Safaricom threatened not to participate in the shared network if it used the 2.6 GHz band, arguing that the 700 MHz band should be used to keep down costs. However, the regulator pointed out that the digital switchover had yet to take place, and hence the band was as yet unavailable. As of May 2012, at which point the government announced that the project would be operational during 2013, the list of probable participants appeared to be Safaricom, Airtel, Essar, Orange, Kenya Data Networks, MTN Business Kenya, Alcatel-Lucent, Nokia Siemens Networks and Epesi Technologies (USA). In January 2013, Essar—the only operator providing solely 2G, announced that it hoped to skip 3G and proceed straight to LTE.

7.1.5 Liberia

In June 2012, Cellcom claimed to have launched 4G, but it is evident that this is in reality a HSPA+ network (TeleGeography 2012p).

7.1.6 Mauritius

Emtel officially launched LTE in May 2012, claiming (not a little unrealistically) a maximum downlink of 150 Mbps (Businessmega 2012). For its part, Orange launched in June 2012 in Port Louis and Bagatelle using the 1,800 MHz band. In the light of the existing poor downlink speeds and high prices for 2G, the response was muted, to say the least (The Media Guru 2012).

7.1.7 Morocco

In May 2012, the regulator announced that it hoped to conduct an auction at some point later in the year with licences to be awarded in 2013. However, this was later set back by 1 year without explanation.

7.1.8 Mozambique

In March 2013, the regulator announced a forthcoming auction of five blocks, each of 5MHz paired, in the 790–862 MHz band. The reserve price was set at $30 million per block (Cellular-news 2013d).

7.1.9 Namibia

In 2010, MTC announced plans to roll out a 4G network but was subsequently frustrated at every turn by an alliance of the regulator, Telecom Namibia and the Windhoek City Council. Despite the absence of a licence which the regulator was in no hurry to issue, MTC conducted trials in the hope of a launch immediate upon gaining a licence. At the end of March 2012, the regulator converted all licences held by MTC, AfricaOnline and Wireless Technologies Namibia (WTN)—but not those of Powercom, CTM or iBurst—onto a technology-neutral basis. In May, MTC launched in Windhoek branded as Netman 4G and committed to covering 45 % of the population within a year (TeleGeography 2012k). In mid-August, MTS announced that it had 1,000 data-only contract subscribers, yet these accounted for 23 % of the total data usage among its more than two million customers (averaging 1.6 GB per day). For its part, Telecom Namibia has contracted with ZTE to construct a 2G/3G/LTE unified network.

The regulator has announced its intention to issue new 1,800 MHz licences earmarked for LTE use but has postponed any decision on digital dividend spectrum due to technical issues.

7.1.10 Nigeria

Currently there are no explicit licences, but existing spectrum can be used—the main incumbents typically hold 5 MHz paired in the 900 MHz band plus 15 MHz paired in the 1,800 MHz band. In March 2012, the regulator announced that it was seeking to reclaim unused spectrum in the 2.5 GHz band from the National Broadcasting Commission which also controls the 700 MHz band. However, given that in the former case it had been trying to reclaim for 4 years, progress may not be rapid (TeleGeography 2012e).

In August 2012, several somewhat unsuccessful CDMA operators announced their intention to merge and then switch over to LTE using their existing holdings of spectrum in the 1,900 MHz band. Whether they will receive regulatory permission to do so is another matter.

7.1.11 Rwanda

In March 2013, KT Telecom signed a MoU with the government of Rwanda establishing a joint venture to provide KT Telecom's LTE technology to the national incumbents via a common network and MVNO arrangements (Cellular-news 2013b). The network is expected to encompass 95 % of the population within 3 years.

7.1.12 South Africa

Vodacom claimed to have conducted the first trial in Africa, which commenced in June 2010 using its 3G spectrum. Cell C has conducted trials in the 850/900 MHz bands while MTN has conducted trials in the 1,800 MHz band, originally with a view to a commercial launch at the end of 2011. However, this proved to be unduly optimistic since there was too little spare capacity for LTE to be launched using existing spectrum and the regulator had yet to auction vacant 2.6 GHz and 3.5 GHz spectrum despite initially pencilling this in for July 2010.

Impatient in the face of the regulator's equivocation, Wireless Business Solutions, the parent of local wireless broadband provider iBurst, announced in October 2011 that it would construct a LTE network and launch it by mid-2012. This would utilise its existing spectrum in the 1,800 MHz and 2.6 GHz bands (TeleGeography 2011c)—however, by July 2013 it had managed only to construct a trial network in the 2.6 GHz band and was looking to form a joint venture with MTN or Vodacom. Another fixed-wireless operator, Neotel, which was awarded spectrum in the 800 MHz band in March 2007 that could be used for LTE, and which also currently provides WiMAX in the 3.5 GHz band, is testing LTE in the 1,800 band (TeleGeography 2012ad). It announced in May 2013 that it intended to launch, initially in the Gauteng region, between July and September that year.

In mid-December 2011, the regulator published its proposals for the 800 MHz and 2.6 GHz bands, based upon a beauty contest due to take place in March 2012. The available spectrum consisted primarily of two blocks of 15 MHz paired, two blocks of 20 MHz paired and two blocks of 20 MHz unpaired in the 2.6 GHz band. The licence conditions included a requirement to host MVNOs and 70 % geographic coverage within 5 years—for 2.6 GHz-only licensees this would be shortened to 4 years. But the potential stumbling point was a stipulation that applicants must be locally licensed and 30 % owned by historically disadvantaged individuals, a stipulation that would, for example, disqualify Vodacom. Indeed, the proposal effectively assigned licences to state-owned broadcast and broadband infrastructure company Sentech and to Neotel. In the event, the auction was postponed until an unspecified future date (TeleGeography 2012c) after a wide range of protests was tabled.

The postponement produced a sharp riposte from MTN which claimed that it was ready to launch as soon as it received a licence, but that it was being starved of appropriate spectrum for 4G—it needed at least 10 MHz paired in the 2.6 GHz band. It pointed out that Sentech held 50 MHz of unused spectrum in this band and that it made no sense to allow Sentech or anyone else in a similar situation to be eligible to bid for additional spectrum. In August, it added that it had switched on 200 LTE test sites and was running a trial using 10 MHz of re-farmed 1,800 MHz spectrum (TeleGeography 2012t). In September, it announced that it expected to have between 400 and 500 LTE-equipped base stations by the year-end, by which point it would have launched in Johannesburg, Pretoria and Durban. In practice, it had erected 1,600 base stations by the time of its December launch (TeleGeography 2012ak).

In October 2012, Telkom (branded as 8ta) announced that it would be conducting a free trial of LTE for customers in the Gauteng area between November 2012 and March 2013 prior to a launch in April. In this respect it was much slower off the mark than Vodacom, which launched on a modest scale in Johannesburg in October 2012 using existing spectrum (Cellular-news 2012d). Pretoria and Durban came next with Cape Town the final large city to be served during 2012. For its part, Cell C announced in November that it would be launching in Cape Town and Durban before the year-end, followed by Johannesburg and Pretoria in January 2013. However, it referred to this as a trial network pending a decision by the government about a national wholesale network (Cellular-news 2012e) and in May 2013 it confirmed that although it was selling a LTE-enabled iPhone 5 it was not as yet providing a LTE signal.

In a move that is certain to be welcomed by the incumbents, state-owned Sentech, which had been awarded 50 MHz in the 2.6 GHz band and 14 MHz in the 3.5 GHz band with a view to it rolling out a WiMAX network, decided that the technical difficulties that it faced were insuperable and agreed in April 2013 to return the spectrum to the regulator.

7.1.13 Tanzania

Smile Communications conducted a soft launch of LTE in Dar es Salaam in June 2012 using the 800 MHz band, with the intention of providing a nationwide service by the end of the year (TeleGeography 2012n), although 2014 is more realistic. It gained roughly 500 customers in its first 2 months of operation but did not announce that it had launched commercially until May 2013 at which point it opened up a wholesale service for other operators. For its part, Vodacom started trials in February while awaiting the allocation of spectrum suitable for a nationwide LTE launch.

In October, Alcatel-Lucent announced that it had been contracted to roll out a network for Telesis, a Mobile Virtual Network Aggregator, but nothing further is yet known.

7.1.14 Uganda

Smile Communications—see also DRC and Tanzania—claimed to have 200 LTE customers in Kampala in November 2012, and announced its intention to cover the whole of the capital by end-March 2013 followed by a full nationwide roll-out. However, Alcatel-Luent, the vendor, only announced the placing of the contract in

November so it was not clear whether a commercial launch had actually taken place (TeleGeography 2012ag). The official launch was announced in June 2013, using the 800 MHz band in Kampala but only providing an average downlink of 6 Mbps (TeleGeography 2013y). For its part, MTN launched at the end of April 2013 in Kampala, claiming that this was the first LTE network to go live (TeleGeography 2013u).

7.1.15   Zambia

The mobile market is dominated by Bharti Airtel. In March 2013, Zamtel applied to the regulator for a LTE licence. In June, start-up Massnet Innovation Solutions announced that it was seeking an investor to help roll-out a LTE network using 2.6 GHz spectrum awarded to it in August 2012 (TeleGeography 2013z).

7.1.16   Zimbabwe

In March 2013, Aquiva Wireless—which is not an incumbent mobile operator—announced that it intended to launch a LTE network. In May, the government proposed legislation that would make infrastructure sharing mandatory. A suspicion arose that this was intended to provide protection for struggling state-owned Net*One (Cellular-news 2013f).

In practice, no specific licence is required for LTE, and with a major United Nations conference due to take place in Victoria Falls at the end of August there is an incentive for Econet to make good on its promise to have LTE up and running there in time for its opening (TeleGeography 2013ac).

7.2   The Middle East

7.2.1   Azerbaijan

Azerfon announced in April 2012 that it would have its first LTE base station up and running in time for the Eurovision Song Contest scheduled to take place there in May and declared that it intended eventually to go nationwide, subsequently amending that to a launch on a nationwide basis. TeliaSonera indicated that it hoped to follow suit—it duly launched in Baku using its 1,800 MHz spectrum (Telecom.paper 2012a)—while Bakcell's launch will come sometime during 2013. However, the status of new spectrum licences is unclear.

7.2.2   Bahrain

Progress in Bahrain has been delayed by the failure to auction dedicated spectrum for 4G—although, unusually, the regulator refers not to 4G or LTE but to 'post-3G

Table 7.1 Auction proposal. Bahrain. 2013

Lot	Licences	Band	MHz	Reserve $m
1	1	900	5.6 paired	4.35
2	1	1,800	10 paired	3.88
3	1	1,800	5 paired	1.94
4	4	2,100	5 paired	1.03
5	2	2,600	20 paired	2.06
6	3	2,600	10 paired	1.03
Total	12		110.6	

services'. Zain commenced its upgrade to LTE at the end of 2009, while Batelco completed testing in January 2012 and STC claimed to have launched in January 2012 (although it indicated that there would also be a 'commercial' launch in April).

In early March 2012, the regulator invited interest in 55 MHz of unallocated spectrum in the 1,900–2,170 MHz band, comprising 20 MHz paired and 15 MHz unpaired specifically for post-3G usage. This would initially be assigned to the three incumbents for a period of 6 months and then put out for auction together with additional spectrum in the band, either at the end of 2012 or—as transpired in practice—in early 2013 (TeleGeography 2012b). Later in the month further spectrum to be sold was announced as an initial tranche of 40 MHz followed by a further 30 MHz, both in the 2.6 GHz band; 15 MHz paired in the 1,800 MHz band; 15 MHz unpaired in the 1,900 MHz band; and 5.6 MHz paired in the 900 MHz band (Cellularnews 2012a). It was subsequently announced that all of the 2.6 GHz spectrum would be sold at the same time even though 30 MHz would not be released until 2014, only for there to be a further announcement in November declaring that all of the spectrum would in fact be issued immediately after the licences—all 15-year and technology-neutral—were awarded (TeleGeography 2012ah).

In January 2013, the initial set of minimum prices was adjusted downwards in respect of the 1,800 MHz licences (by 25 %) and the 2,600 MHz licences (by 20 %). Furthermore, the unpaired block in the 1,900 MHz band was dropped due to lack of interest. This then left the position as shown in Table 7.1 (TeleGeography 2013e).

However, no licensee would be permitted to exceed a cap of 90 MHz paired totalled across all of its licensed bands including any pre-existing spectrum. Pre-existing holdings were as follows:

- Batelco—11.8 MHz paired/900 MHz; 20 MHz paired/1,800 MHz; 15 MHz paired/2,100 MHz. Total = 46.8 MHz paired.
- Zain—12 MHz paired/900 MHz; 25 MHz paired/1,800 MHz; 10 MHz paired/2,100 MHz. Total = 47 MHz paired.
- Viva—5.6 MHz paired/900 MHz; 15 MHz paired/1,800 MHz; 15 MHz paired/2,100 MHz. Total = 35.6 MHz paired.

Hence, Batelco could acquire up to 43.2 MHz paired, Zain up to 43 MHz paired and Viva up to 54.4 MHz paired. Coverage obligations would apply to any bidder winning more than 20 MHz paired in the 2,600 MHz band or winning any spectrum at all in the 900 MHz and 1,800 MHz bands. The coverage, set at 99 % of the population, would have to be achieved within 9 months by incumbents or 2 years by new entrants.

In February 2013, Batelco announced the launch of the 'first real 4G LTE' network in Bahrain, using the 1,800 MHz band but with unspecified coverage (TeleGeography 2013g), which implicitly cast aspersions on prior claims made by STC. It certainly did not seem to be the case, judging by statements made by STC to the effect that 'in August 2012, Viva offered a demonstration LTE network to the Bahraini public'—see the Company Overview at http://www.telegeography.com—that it had actually conducted a launch prior to that of Batelco. Zain launched in late April and it was indicated at that time that it was the second launch (TeleGeography 2013o).

In November 2012, a WiMAX operator with a nationwide network using the 3.5 GHz band, Menatelecom, announced that it intended to be the first operator in the world to launch a nationwide LTE network (TeleGeography 2012al) and signed a MoU with Huawei in December—although it has to be said that Bahrain is quite small in area as nations go. However, it was precluded from participation in the forthcoming auction which was restricted to the three incumbents (TeleGeography 2013l), a decision it promptly appealed on the grounds that it had previously received a licence to operate a fully-mobile network. After the appeal was won, the regulator promptly filed a counter-appeal, but this was counter-balanced by a further case being set in hand by 2Connect on the grounds that it had also been barred from bidding (TeleGeography 2013p). In June 2013, the licence was cancelled.

7.2.3 Belarus

Turkcell (BeST) and MTS Belarus (majority owned by the state) completed trials pending a 2.6 GHz band auction. However, the market looked to be developing similarly to Russia in that Scartel (branded as Yota) set out to use its 30 MHz paired of spectrum in the 2.5/2.6 GHz bands, licensed in December 2008, to build out a LTE network rather than the WiMAX network originally envisaged. It soft-launched initially in Minsk and Grodno in December 2011 and the commercial launch followed in May 2012 (TeleGeography 2012i). The operators were expected to re-sell services rather than build out separate networks at great cost, but the announcement in June by Scartel that it would be closing its network in order to concentrate upon Russia has left things in a state of confusion.

In June, BelCel announced that it was preparing to test either LTE or cdma2000 1xEV-DO Rev. B in the 450 MHz band. Little progress was made until late April 2013 when the government surprisingly announced that tests were shortly to begin using LTE-Advanced with the tests conducted by MTS and local cloud technology company becloud (TeleGeography 2013r).

7.2.4   Bosnia and Herzegovina

Telekom Srpske, a subsidiary of Telekom Srbija, conducted a successful trial in May 2013. For its part, state-owned BH Telekom contracted with Ericsson and Nokia Siemens Networks to roll out a trial network.

7.2.5   Iraq

In February 2013, Alcatel-Lucent announced that it would be rolling out a LTE network for Regional Telecom which held a regional licence in Kurdistan. No details of any kind, including the origins of the 4G licence, were made available (Cellular-news 2013a), possibly because the relevant regulator is located in Kurdistan. The launch took place in June.

7.2.6   Israel

In January 2013, the government announced that it would be authorising the use of the 1,800 MHz band for LTE. Since Cellcom and Partner Communications already owned spectrum in this band, new spectrum would be allocated only to Pelephone, HOT Mobile and Golan Telecom and only then provided they intended to roll out a network. Steps would be taken to ensure that Cellcom and Partner could not launch much earlier than their rivals (Telecom.paper 2013a).

In May 2013, the government announced that the defence forces had now vacated the 700 MHz and 2.6 GHz bands but were holding on to the licences until such time as they received the agreed amount of compensation. There were also residual issues in relation to Bezeq which provided fixed-wireless services within these bands. As a result, an auction is unlikely to take place before the end of 2013, and re-farming is impractical because the existing 2G/3G networks have no spare capacity (Cellular-news 2013e).

7.2.7   Jordan

In December 2012, it was announced that an auction would be held in early 2013 for spectrum in multiple bands: 1,800 MHz ($46 million minimum per 5 MHz paired for 2G, 3G or 4G); 2.1 GHz ($35 million minimum per 5 MHz paired for 3G); 2.3 GHz ($15 million minimum per 5 MHz for 4G and $3 million minimum per 5 MHz for WLL); and 2.6 GHz ($26 million minimum per 5 MHz paired for 4G) (Telecom.paper 2012f). One new mobile operator would potentially be licensed.

In March, Umniah Telecommunications completed a technical test of LTE but, given that it had recently introduced DC-HSPA+, it indicated that it was in no hurry to launch LTE, a sentiment echoed by Orange (TeleGeography 2013j).

In practice, the auction process was not set in hand until June 2013 with bidding due to close in September. In response, the incumbents accused the authorities of using the sector as a cash cow via a proposal to raise income tax on the sector from 25 to 40 %, and threatened to boycott the auction.

7.2.8 Kazakhstan

The first mention of LTE in this country appears to be a statement by VimpelCom that it was testing in the 700–800 MHz band, one of three—the others were the 2.3 GHz and 2.6 GHz bands—seemingly deemed suitable for LTE. However, the regulator responded that this had not been authorised. In May 2012, the Minister of Communications announced that the first commercial launches would take place in Almaty and Astana in 2013, and claimed that all towns with a population of over 50,000 would have access to LTE before the end of 2015 and the rest of the country before the end of 2018.

Curiously, only state-owned fixed-wire operator Kazakhtelecom (KT), which is a minor player in the mobile market via CDMA operator Altel, was named by the Minister, leading analysts to conclude that KT would build the only network and operate as a wholesaler. However, little is certain at this point in time (TeleGeography 2012m) although it would appear KT launched a trial followed by a commercial launch in Astana and Almaty in December 2012—the spectrum band was not specified (Telecom.paper 2012i). KT claimed to have 30,000 subscribers at the end of March 2013, providing them with a maximum downlink of 30 Mbps (TeleGeography 2013s).

TeliaSonera (K'Cell) is another incumbent which can be expected to become heavily involved in the provision of LTE, to which end it has acquired spectrum in the 2.6 GHz band via the purchase of WiMax assets in six cities from Alem Communications.

7.2.9 Kuwait

According to one report (LteWorld 2012) and its own website, Viva launched a commercial service in December 2011. However, in October 2012, the government authorised the use of the 1,800 MHz band re-farmed for LTE, offering licence amendments to the three incumbents for $888,000 apiece (TeleGeography 2012z), so it is unclear what spectrum Viva could have used for its prior launch—its website is silent on the matter. In November, Zain launched a nationwide service (Cellular-news 2012f) branded as 'Wiyana Connect 4G LTE', whereas Qatar Telecom (al-Wataniya) delayed until July 2013.

Saudi Telecom (STC) signed a roaming agreement with Viva, of which it is a minority shareholder, in December 2012. In March 2013, Viva stated that its LTE service was now available to all of its customers. LTE appears to be (too big?) a success given that during 2013H1, for example, Zain managed to shift 10 % of its customers over to LTE and these accounted for 40 % of its data traffic.

7.2.10 Kyrgyzstan

In October 2012, the regulator awarded LTE/WiMAX licences—details were not made available—to 12 companies (spellings are variable): Foris Telecom, Saima Telecom, Global Asia Telecom, ToTel, Aknet, Kurulush Invest, T-Com, Fraton Plus, Aytel, WTT, Asia Info and Intranet KG (Telecom.paper 2012c). So far, only the first four appear to have commenced the roll-out of their networks and Saima is the only one to have launched—using dongles and the 2.6 GHz band in Bishkek, the capital city, in December 2011 (Barton 2012).

7.2.11 Lebanon

State-owned Alfa launched LTE in Central Beirut in May 2013 but no handsets were available at the time. It expects to achieve 40 % population coverage by the year-end (TeleGeography 2013w). MTC (Touch) the other state-owned operator—both are privately-managed—followed suit within a matter of days (TeleGeography 2013x).

7.2.12 Moldova

In August 2012, the regulator announced that the three incumbents—TeliaSonera (Moldcell), Orange and Moldtelecom (Unité) would be offered 15-year technology-neutral licences, each of 20 MHz paired in the 2,500–2,560 MHz and 2,620–2,680 MHz bands, at the year-end for €10 million ($12.3 million) apiece. Each licence would also come packaged with 50 MHz in the 3,600–3,750 MHz band. Should any licences be declined, they would be offered via an auction together with a further licence in the 3,750–3,800 MHz band costing a minimum €1 million (TeleGeography 2012r). TeliaSonera and Orange received their licences in early November and both launched almost immediately (Telecom.paper 2012e).

It must be presumed that Moldtelecom did not receive a licence since, in April 2013, the regulator announced that it would be holding a beauty contest for a technology-neutral licence in the 2,500–2,520 MHz paired with 2,620–2,640 MHz band. The winner would be obliged to provide 4G services to 25 % of the population (excluding Transnistria) by end-2015 (TeleGeography 2013q).

7.2.13 Oman

As things stand, it is not possible to be certain what is happening. Having conducted successful tests in July 2010, Omantel constructed a TD-LTE network using equipment supplied by Ericsson and Huawei. However, the regulator intervened in August to make it clear that the 30 MHz of spectrum in the 2.3 GHz band that it had awarded to Qatar Telecom (Qtel, trading as Nawras) and Omantel was exclusively for fixed-wire connectivity. In July 2012, Omantel announced a launch without itself specifying the spectrum used but which is claimed by other parties to have been the 2.3 GHz band (Cellular-news 2012b), in which case it cannot be a mobile network. Omantel is also building a FDD network in the 1,800 MHz band as is Qtel which had initially conducted its own trial of TD-LTE in March 2012. This was reported to have been launched in August (using the brand Ooredoo), but the official announcement of a commercial launch in Muscat did not transpire until February 2013 (TeleGeography 2013h).

7.2.14 Qatar

In December 2010, Vodafone carried out tests using the 800 MHz band. For its part, Qatar Telecom (Qtel) carried out tests in the summer of 2011, also using the 800 MHz band, and in October it announced that it would be building 900 LTE base stations. However, the 800 MHz band is in use for analogue TV and there is no timetable to clear the band and auction it for LTE use, so in the interim the mobile incumbents will be obliged to use their existing bandwidth (TeleGeography 2012d). In September 2012, it was announced that Qtel had received spectrum in the 800 MHz and 2.6 GHz bands, that it was conducting tests in Doha and that it expected to launch commercially before the year-end. However, by January 2013 this ambition had lessened somewhat to an unspecific desire to launch in Doha during 2013 at speeds roughly three times as fast as the existing service (TeleGeography 2013d). For its part, Vodafone expressed the intention to launch somewhat earlier than Qtel, but in the event Qtel (now branded as Ooredoo) was the first to launch in April 2013 using the 800 MHz and 2.6 GHz bands in Doha (Cellular-news 2013c).

7.2.15 Saudi Arabia

On 14 September 2011, all three mobile operators—Saudi Telecom (STC), Zain and Etisalat (Mobily)—claimed to have been the first to launch. However, none provided the level of detail about their services which would have made it possible to verify which was first to market (Lennighan 2011). In January 2012, Etisalat— which launched TD-LTE using the 2,570–2,620 MHz band (Band 38)—admitted that it had insufficient spectrum to roll out LTE on a nationwide basis while, in June, Zain announced that it intended to re-farm its 1,800 MHz spectrum in order to boost its capacity for 4G services. For its part, STC announced in May that TD-LTE was now available in 38 cities using the 2.3 GHz band (Band 40)—although the commercial launch was in September—and that it expected to achieve 65 % population coverage by the year-end (TeleGeography 2012l), somewhat less than the 85 % coverage claimed by Etisalat in November when it revealed a roaming agreement with Hong Kong's CSL. STC signed a roaming agreement with Kuwait's Viva, of which it is a minority shareholder, in December 2012.

The combination of three apparently simultaneous launches, of FDD (Zain) and TDD (Etisalat and STC) and of three different spectrum bands (1,800 MHz, 2.3 GHz and 2.6 GHz) is unlikely to be repeated elsewhere.

7.2.16 Turkey

Turkcell began tests in May 2010 and by mid-2012 had proved itself capable of delivering a 100 Mbps downlink. In April 2013, the government stated that a joint venture involving Asiesan Elekronik and Ticaret, Netas Telkomunikasyon and Turk Telecom's software subsidiary Argeia had contracted with the government to build a LTE network capable of a maximum downlink of 100 Mbps for civilian and military use (TeleGeography 2013n).

7.2.17 Ukraine

MTS is conducting tests while VimpelCom, post-restructuring, has spare spectrum for LTE use. In September 2012, in response to lobbying by Private Group, parent of cdma2000 operators PEOPLEnet (TeleSystems) and ITC (CDMA UA)—plus, indirectly, the latter's subsidiary Velton Telecom—which were considering a switch to LTE, the government authorised the use of the 1,900–1,920 MHz and 1,980–2,000 MHz bands for high-speed data use.

In October, all CDMA 800 MHz operators with regional licences were authorised to re-farm their spectrum with a view to collaborating in an attempt to optimise joint geographical reach and capacity. The operators involved comprised Astelit (which also possessed GSM licences), ITC/Velton, Intertelecom and Ukrainian Wave (part-owned by Ukrtelecom). Subsequently, subject to the issue of technology-neutral licences, the 'redesigned' allocation of spectrum would be opened up for LTE. In late October, Astelit gained 5 MHz in each of the 15 regions not yet licensed via the re-farming process, thereby increasing its coverage to all 27 regions. Meanwhile, Intertelecom, which already served all 27 regions with cdma2000 1xEV-DO, was able to increase its bandwidth to 17.5 MHz in 20 regions and at least 10.5 MHz elsewhere, while ITC/Velton appeared to be unaffected—the latter already provided cdma2000 1xEV-DO Rev. A in 26 regions compared to the 27 regions served by PEOPLEnet. Infrastructure-sharing arrangements are under negotiation.

In November 2012, the regulator awarded Ukrtelecom a nationwide licence in the 5,250–5,350 MHz band, with the full 100 MHz made available in all bar one region. Services are expected to come on stream in June 2013 (TeleGeography 2012ac). It also awarded a nationwide licence in the 1,980–2,000 MHz band to First Investment Alliance (TeleGeography 2012ai).

In March, the regulator announced that all incumbents should have a 3G licence by the year-end, finally putting an end to the prevarication of the military. But the Ukraine remains a clear laggard in relation to LTE.

7.2.18 UAE

Du completed the roll-out of its network in December 2011, but due to the absence of devices that were compatible with its use of the 1,800 MHz band, chose not to launch until July 2012 when it could serve 28 % of the population. It subsequently conducted successful tests of LTE using 4×4 MIMO, achieving a maximum downlink of 300 Mbps. Meanwhile, the other incumbent, Etisalat, had launched in the majority of urban areas in 2011 using a combination of spectrum in the 1,800 MHz and 2.6 GHz bands (TeleGeography 2012o), achieving 80 % population coverage by end-September 2012.

In May 2013, the regulator announced its intention to combine the 800 MHz band plan in use in Europe with 30 MHz paired in the lower section of the 700 MHz band as used by the APT.

7.2.19 Uzbekistan

TeliaSonera (UCell) was licensed to provide LTE using the 2.6 GHz band as early as July 2010 and launched in central Tashkent the following month. However, it lagged MTS which was licensed in the same band in October 2009 and which launched in central Tashkent in July 2010, claiming this to be the first LTE launch in the CIS and Central Asia (TeleGeography 2010). The third main incumbent, VimpelCom (Unitel), did not launch until December 2011.

The situation in respect of MTS then became mired in controversy, with the result that its licence was revoked in August 2012 on the basis of alleged criminal activities and failure to pay back-taxes and it entered insolvency proceedings in January 2013.

7.3   The Americas

7.3.1 Antigua and Barbuda

The government of Antigua and Barbuda is engaged with other Caribbean states in forming a harmonised approach to a digital switchover by 2015. Meanwhile, it set out to issue two LTE licences in 2013Q1 (TeleGeography 2012u). Given that it has expressed its displeasure with incumbents Digicel and C&W (LIME) on account of their reluctance to proceed with LTE, the government may use 4G licensing as an opportunity to introduce a third operator into the market. However, Digicel has stolen a march on potential rivals by launching in November 2012 (see http://www.digicelantiguaandbarbuda.com).

7.3.2 Argentina

The incumbents were all facing capacity constraints so they were all anxious to acquire new spectrum for 3G and/or 4G use. In September 2012, the government cancelled the much postponed sale of 3G licences in the 850 MHz and 1,900 MHz bands, comprising spectrum returned by Telefónica and originally put on offer in May 2011, after a beauty contest in which América Móvil, Telecom Personal,

NII Holdings, Supercanal and Multitrunk had submitted initial offers for spectrum in the 1,900 MHz band. It then unilaterally assigned the spectrum—either 30 MHz or 35 MHz depending upon the region—together with 7.5 MHz in the 850 MHz band, to state-owned Arsat (Artel). Although it was alleged that (all bar one of) the private bids failed to meet the conditions of the tender, the decision appeared to be politically influenced—it was noted that, unlike the four incumbents, Arsat (trading as Libre. com) would need to build its network from scratch (Wireless Federation 2011). Subsequently, China's Datang Mobile Communications indicated a desire to co-operate with Arsat which also seems likely to be awarded a significant chunk of the spectrum available for 4G.

Telefónica and Telecom Personal had conducted LTE trials during late 2010 using existing spectrum, but subsequently were constrained by the uncertainty over who would end up with what spectrum. An auction of spectrum intended specifically for 4G use is expected to take place in 2013.

7.3.3 Bahamas

BTC, the incumbent, is undertaking an upgrade of its network to HSPA+ which it is marketing as 4G (TeleGeography 2012f). However, LTE is on the way using 700 MHz spectrum auctioned in September 2012. 72 MHz was put on offer via 15-year licences, with an individual operator capped at 24 MHz, and five companies applied of which two were rejected (Cellular-news 2012c). BTC received the first licence, comprising a 24 MHz block, in early September which came with population coverage requirements of 75 % on the four main islands within 18 months, rising to 99 % within 30 months plus further, less-onerous requirements on other islands (TeleGeography 2012w). BTC intends to commence its roll-out in January 2013 and to launch some 3 months later. Cable Bahamas Ltd is to be awarded the second licence which will run from when BTS loses its monopoly in April 2014.

7.3.4 Bolivia

In the near future, the government intends to auction spectrum in the 1,500–1,600 MHz, 1,900 MHz and AWS bands on a nationwide basis. In addition, spectrum in the 900 MHz, 2.6 GHz and 5.2 GHz bands will be on offer with more limited geographic coverage (TeleGeography 2013a).

However, in December 2012, Entel claimed to have launched in the lower section of the 700 MHz band as used in the USA, offering up to a 40 Mbps downlink in La Paz, Cochabamba and Santa Cruz (Prescott 2012). At $101 per month via a modem, this seems unlikely to attract many users, but the privately-owned incumbents will have quite a lot of catching up to do.

7.3.5 Brazil

Nothing concrete was apparent before 2012 although operators buying a MMDS 2.6 GHz spectrum owner were able to launch prior to the issue of formal 4G licences. In December 2011, after selling off 15 small blocks in the 800 MHz and 1,800 MHz bands, the regulator announced that it would be auctioning 120 blocks of unspecified spectrum in the 2.6 GHz and 450 MHz bands—the latter for rural coverage—at some point during 2012 which was subsequently pencilled in as June. The objective was stated to be the provision of 4G services to all state capitals and all cities with more than 500,000 inhabitants by 2014. Altogether, the auction of 140 MHz of spectrum would include four nationwide licences, with two providing 20 MHz paired and two providing 10 MHz paired in the 2.6 GHz band. The licensees would be required to provide a 450 MHz band service in rural areas— where a rural area is defined as one at least 30 km from any municipality—so would be expected to bid for this spectrum in the parallel auction (Lennighan 2012).

Both Telecom Italia (TIM) and Telefónica (Vivo) expressed severe misgivings about holding an auction as early as 2012 as they considered that the market was still trying to come to terms with 3G, and a legal challenge was considered. However, the four main incumbents duly entered the auction with Telefónica, América Móvil (Claro), Telecom Italia and Oi paying respectively $509 million for 20 MHz paired, $409 million for 20 MHz paired, $183 million for 10 MHz paired plus some regional licences and $168 million for 10 MHz paired plus some regional licences (Rayal 2012). No bids were entered for spectrum in the 450 MHz band, so these four licensees will be forced to invest in rural roll-outs in this band in designated regions as part of their 2.6 GHz licences. Specifically, 30 % of an operator's rural areas will have to be covered by June 2014, 60 % by December 2014 and all by December 2015. In comparison, the 2.6 GHz licensees will initially be required to launch services in the six cities hosting football's Confederations Cup by April 2013 and in the 12 cities hosting the 2014 Football World Cup by the end of 2013. The first handset on the market was the Motorola RAZR HD in September 2012.

América Móvil set in hand a series of trials with Ericsson during 2012 with a view to a launch in April 2013, while Oi began trials with Nokia Siemens Networks in September 2012 with a view to providing a service in seven cities by the same date. This was achieved in Rio but not elsewhere, with Oi arguing that it was only required to have the technology in place and not to have launched commercially by the deadline. It signed an infrastructure sharing agreement with Telecom Italia (TeleGeography 2013t) which was ratified by the regulator in mid-April. For its part, Telefónica announced that it would launch in four cities in April 2013 using the brand '4G+'. In practice, América Móvil was the first to launch, using the '4GMax' brand in Recife in December 2012 (Telecom.paper 2012h), following on with what presumably was its full commercial launch in Recife (where it had 15,000 subscribers) and a further ten cities in April 2013 (Telecom.paper 2013d). It followed this in May by signing an infrastructure sharing agreement with Telefónica in May.

Two hundred sixty-eight lots of WLL spectrum in specified regions were also auctioned, with some going to the mobile incumbents as noted above. In addition, Sunrise proved successful in São Paulo at a cost of $9 million and Sky Brasil in São Paulo, Rio de Janeiro and elsewhere at a cost of $44 million, but by no means all of the lots were sold. Strictly speaking, this did not appear to be the first step in fixed-wireless provision since, in September 2011, Sky Brasil had signed a deal with fixed-wire operator Telebras to begin offering services in 2012, and had launched a 2.6 GHz band TD-LTE service in Brasilia in December 2011 via a Wi-Fi router.

Although some spectrum in the 3.5 GHz band will become available during 2012, the digital dividend band switchover is not due to take place until July 2016, in the meantime creating a situation of excess demand for spectrum suitable for high-speed data services. However, the regulator now seems intent upon wresting back spectrum in the 746–806 MHz band from broadcasters in order to be able to auction it off during 2014 (TeleGeography 2013f). In February 2013, the government announced that it would be using the APT band plan, adding in June that the analogue switch-off would be brought forward in larger cities from 2016 to early 2015.

In October 2012, TIM announced that it would be deploying LTE in the 450 MHz band in order to improve rural coverage, but would be using the 2.6 GHz band to provide coverage in advance of the 2013 Confederations Cup. In November, Telefónica signed a MoU with Sweden's AINMT with the same intention in mind. For its part, regional operator ON stated that it would be launching an Air4G LTE-A network immediate upon completion of its roll-out in March 2013 providing the relevant licence had been secured.

In February 2013, the regulator gave an ultimatum to the multichannel multipoint distribution service (MMDS) pay-TV providers to the effect that they had to vacate the 2.5 GHz and 2.6 GHz bands by 12 April so that they could be re-farmed for LTE. The problem was that there was supposed to be a LTE network operational in cities where the Confederations Cup was to be played in the summer, so the regulator threatened sanctions if the spectrum was not cleared in time. In May, the regulator ruled that the four incumbents would be obliged to pay $157 million to MMDS operators, with Telefónica, América Móvil and Sky Brasil receiving most of the money.

Meanwhile, in March, Telefónica announced an infrastructure-sharing agreement with América Móvil which set out to include spectrum sharing, while TIM made an arrangement (the so-called RAN sharing LTE project) with Oi that was limited to physical infrastructure and the 2.6 GHz band (TeleGeography 2013i). TIM launched in Sao Paolo in July. For its part, the Center for Research and Development (CPqD) presented a network solution for the 450 MHz band which would provide a 25 Mbps downlink and 12.5 Mbps uplink.

7.3.6 British Virgin Islands/Cayman Islands

Cable and Wireless (LIME) has nominally launched in both sets of islands. However, it is evident that in both cases the '4G' advertised is in fact HSPA+

(TeleGeography 2012q). In December 2012, Digicel also launched HSPA+ in the BVI—and even then only at a maximum downlink of 21.1 Mbps—while claiming to be providing '4G'.

7.3.7   Canada

Canada freed up the digital dividend 700 MHz (700–800 MHz) band in August 2011, prior to which, at the end of March, it converted fixed-wireless licences in the 2.6 GHz band to technology-neutral Broadband Radio Service licences. Pending this, Rogers Communications—which together with Telus and Bell, the other national incumbents, control over 90 % of the market—began a set of trials in October 2010 using existing 3G spectrum, and it also hoped to start trials using the 700 MHz band. A launch in the biggest urban centres was pencilled in for late 2011 but took place as early as July in Ottawa using the 2,100 MHz band—it also launched with its budget brand subsidiary 'Fido' in July 2012. Joint ventures such as Inukshuk Wireless (a 2.6 GHz band operator jointly-owned by Bell and Rogers) were expected to be used to exploit the 2.6 GHz spectrum, but Inukshuk was disbanded in March 2012. Rogers announced in November that it was now possible for any customer buying a special version of the LG Optimus G handset to access 4G using the 2.6 GHz band in every market where access was already possible using the 2,100 MHz band.

Bell, having expressed the view that there was unlikely to be a launch before 2012, proceeded to launch in Ottawa in September 2011 using dongles. In November, it launched its first handset, the HTC Raider 4G and had reached 14 cities by February 2012. By November, it had introduced a service using the 2.6 GHz band in just a few large cities (TeleGeography 2012ab). For its part, Telus launched in 14 cities in February 2012 using AWS 1,700/2,100 MHz spectrum while awaiting access to the 700 MHz band for rural provision.

In March 2012, the government announced that it would be removing the ownership restrictions that existed in respect of foreign investors in mobile operators with less than a ten per cent share of total market revenue in order to open up auctions of spectrum suitable for 4G. In addition, the larger domestic operators would be subjected to a spectrum cap, but there would not be any licences reserved for the smaller operators as there had been in the 2008 AWS auction on the grounds that the new entrants in 2008 had failed to provide serious competition for the incumbents.

In January 2013, Rogers announced that, subject to regulatory approval, it had agreed to purchase Shaw's unused AWS spectrum, consisting of 18 regional licences issued in September 2009, in 2014. It added that it had instigated international roaming with Hong Kong. For its part, Telus claimed that its network now covered nearly 70 % of the population and in May it provisionally took over regional operator Mobility subject to regulatory approval which was denied in June—Telus responded by suing the government in July (TeleGeography 2013af). In contrast, Rogers stated that it would have coverage of 95 markets by the year-end using both AWS and 2.6 GHz band spectrum and that in most cases the theoretical maximum downlink would be 150 Mbps.

In March, the government announced that the 700 MHz auction would be held in November—subsequently put back to January 2014. Bell, Rogers and Telus would be subjected to spectrum caps in order to ensure that at least four operators were present in each region. The auction method would switch from the previous SMRA method to a Combinatorial Clock with bids being tabled for packages of regional licences. All told, a further 68 MHz of spectrum would be added to the 460 MHz already available—a further 60–120 MHz, depending upon the region, is to be auctioned off in the 2.6 GHz band in 2014. Beyond that it is a matter of clearing bandwidth which is already occupied or where there are significant interference issues.

In May, Rogers and regional operator Videotron announced that they had signed a 20-year agreement to operate a shared LTE network in Québec and Ottawa while retaining independent customer bases. The following month Rogers announced a network-sharing and roaming agreement with Manitoba Telecom Services commencing that month. However, the release of the government's spectrum licence transfer framework in June did not favour any attempt by the incumbents to take any actions that diminished competition via spectrum concentration, the objective being to end up with four operators competing in every region. This principle was used to prevent the takeover of Mobilicity by Telus, and Verizon Communications is currently considering whether to take over Mobilicity and, possibly, Wind Mobile as a route into the Canadian market (TeleGeography 2013aa).

It is worthy of note, in relation to the discussion of the prospects for satellite provision in the USA (see Chap. 3), that Dish Network is seeking to establish itself as a provider of LTE services in Canada. In February 2012, the Canadian communications ministry, Industry Canada, approved the transfer to Dish of the Canadian licences held by now bankrupt TerreStar and DBSD North America. If it receives the requisite regulatory approval in the USA, Dish will now be able to launch across the whole of North America (TeleGeography 2012a). However, it faces competition in Canada from Xplornet Communications which in February 2012 launched what it called '4G satellite broadband'. It has obtained all of the Canadian Ka-band capacity on the ViaSat-1 satellite launched in October 2011 and intends to supplement this with capacity on the to-be-launched Hughes-Jupiter satellite. In combination with existing satellite and terrestrial capacity, Xplornet expected to be able to provide a nationwide service by the end of 2012. In July 2012, it announced the successful launch of the EchoStar XVII satellite on which it had acquired all of the Canadian capacity.

7.3.8 Chile

The regulator hoped to award up to four licences, covering up to 140 MHz of spectrum, by mid-2011, in anticipation of which Entel PCS, Telefónica and América Móvil launched trials during 2010. They were able, if they so wished, to launch using their 2G spectrum. In December 2011, the regulator offered 2,505–2,525 MHz, 2,525–2,545 MHz and 2,545–2,565 MHz with a restriction of one block per applicant. Licensees would be obliged to roll out their networks within 1 year of receiving their licences, and would be obliged to launch in 543 specified locations within 2 years. Fifteen companies registered a provisional interest. In May, the regulator initially scored prospective applicants in a beauty contest which saw Entel win with a score of 99.06, América Móvil (given its licence in March 2013) come second with 98.52 and Telefónica come third with 98.11 (TeleGeography 2012j). In July 2012, as the only bidders, they were provisionally awarded Blocks C (at a cost of $8.8 million), A (at a cost of $2.9 million) and B (at a cost of $0.5 million) respectively (TeleGeography 2012s).

In April 2012, Entel PCS began trials using its 100 MHz of spectrum in the 3.5 GHz band while acknowledging that it was not ideal for LTE. For its part, the regulator announced that it was considering a second auction, this time in the 700 MHz band, during June or July 2013. It went on to state in February 2013 that this would be the APT version but using the 703–748 MHz and 758–803 bands. The licences would be nationwide.

In June 2013, América Móvil launched in Santiago and parts of Vina del Mar providing a fairly modest average download of 10 Mbps (Telecom.paper 2013e).

7.3.9 Colombia

Cable TV provider UNE-EPM won 10 blocks, each consisting of 5 MHz of 2.6 GHz band spectrum, in a June 2010 auction at a total cost of $40.8 million (Grupo-Convergencia 2012), and is in partnership with Millicom (Tigo)—plans are well advanced to merge the companies in 2013H2 with UNE-EPM holding 50 % +1 share in the merged entity but with Millicom in operational control. It activated its network in December 2011 but a commercial launch was delayed until June 2012. By November it had coverage in four cities. It is considered to be an incumbent in relation to any further auctions.

In August 2011, 30 MHz of spectrum in the 1,900 MHz band was auctioned off, with Telefónica winning 15 MHz, América Móvil winning 5 MHz and Millicom winning 5 MHz at a total cost of $80 million (Globaltelecomsbusiness 2012).

The Ministry, via Resolution 3263 of December 2011, indicated its intention to auction 225 MHz of AWS (1,710–1,755 MHz paired with 2,110–2,155 MHz, divided into three licences each of 15 MHz paired), 1,900 MHz (1,865–1,867.5 MHz paired with 1,945–1,947.5 MHz at a reserve price of $19.4 million) and 2.6 GHz (2,525–2,570 MHz paired with 2,645–2,690 MHz), divided into three licences each of 15 MHz paired, plus one licence comprising 2,575–2,615 MHz of TDD spectrum. An individual operator would be capped at 85 MHz in the high-frequency bands—the three original incumbents already had 55 MHz and UNE-EPM had 50 MHz—and at 30 MHz in the low-frequency bands (Telecom.paper 2012b), while one licence in the AWS and the unpaired licence in the 2.6 GHz bands would be reserved for new entrants. Initial interest was expressed by the three GSM operators (América Móvil, Telefónica and Millicom) together with Avantel, Nextel and Brazil's Oi. The auction was initially postponed until September 2012 and then again until December.

In November 2012, the auction was postponed once again, this time until February 2013. A successful new entrant in the 2.6 GHz band was now to be given the choice of the 40 MHz of unpaired spectrum or 15 MHz paired at the equivalent price per megahertz. Somewhat unusually, an incumbent would have to pay a surcharge of $49 million per AWS licence won and $34.3 million per 2.6 GHz licence won whereas for a new entrant the equivalent figures were $32.2 million and $7.5 million. A successful bidder would be given 1 year to achieve 150,000 LTE customers if a new entrant or 350,000 customers if an incumbent. By mid-2014, coverage would need to have reached the main cities in all of the 1,123 municipalities (TeleGeography 2012ae).

Meanwhile, in October 2012, it was announced that trials of LTE in the APT version of the 700 MHz band would commence later in the year, with an auction pencilled in for some point during 2013. It was rumoured that América Móvil would not be permitted to bid for new spectrum due to its existing 60 % market share, but in the event it was authorised to bid exclusively in the 2.6 GHz band.

The applicants for the June 2013 auction proved to be América Móvil, Telefónica, Millicom/ETB, Avantel, DirecTV and a consortium led by Mexico's Azteca. Chile's Entel did not apply. In May, Telefónica, Millicom, ETB and UNE-EPM signed an agreement to roll out a jointly-owned network. Subsequently, UNE-EPM announced that it had acquired 85,000 LTE subscribers during its first year of commercial operation and was adding a converged fixed-wireless LTE option for all customers.

In June, it was announced that five licences had been awarded in the AWS and 2.6 GHz bands. América Móvil won 15 MHz paired in the 2.6 GHz band for $62.2 million and DirecTV paid $37.2 million for 15 MHz paired and a further $40.2 million for 40 MHz unpaired in the same band. Telefónica and Millicom/ETB won the unreserved AWS spectrum, paying $102.5 million and $101.4 million respectively whereas Avantel won the reserved spectrum for $55.7 million. Azteca was unsuccessful (MinTIC 2013) and the 1,900 MHz spectrum was left unsold.

7.3.10 Costa Rica

In October 2012, the regulator authorised the takeover of Cable Vision, an ISP, by Grupo ICE on condition that it returned 350 MHz of spectrum in the 1,880–1,920 MHz, 2,520–2,620 MHz, 2,640–2,690 and 3,440–3,600 MHz bands. When these are put up for re-sale they will presumably be opened up for 4G services (TeleGeography 2012aa). Meanwhile, ICE proceeded towards its own launch, achieving this in San Jose in June 2013 using the 2.6 GHz band but only with a dongle. Customers were led to expect a downlink of between 10 Mbps and 20 Mbps.

7.3.11 Dominican Republic

Orange Dominicana claimed to have launched the country's first LTE network in Santo Domingo in July 2012 using the 1,800 MHz band. However, the regulator stepped in during September to prevent Orange from advertising its service as 4G until such time as it could be determined whether it was indeed LTE or a variant of HSPA that was being provided (TeleGeography 2012x). In November, the regulator appeared to accept that LTE was indeed being provided while at the same time arguing that Orange had failed to conduct the proper technical studies prior to launching the network, and hence the suspension of the service—now reported as dating from mid-August—was re-confirmed. The suspension was lifted in January 2013 on condition that only dongles could be used for access and that only the five areas where a launch had so far taken place could be supplied. The permitted spectrum bands were 1,720–1,730 MHz paired with 1,815–1,825 MHz (TeleGeography 2013b).

In March 2013, Tricom also claimed to have launched without specifying the spectrum used and without offering any handsets (Telecom.paper 2013b). However, it was later disclosed that the 800 MHz and 1,900 MHz bands had been used by Huawei to construct a combined cdma2000 1xEV-DO/LTE network which initially covered Santo Domingo and other major cities.

7.3.12 Ecuador

In December 2012, state-owned CNT—a fixed-wire operator with a tiny share of the mobile market—was awarded 30 MHz in the (APT version of the) 700 MHz band and 40 MHz in the AWS band (Telecom.paper 2012g). This specific favouring of a state-owned operator—see also Argentina and Uruguay—appeared to reflect a view in government circles that the private operators were over-charging due to a lack of competition.

7.3.13 Greenland

In July 2013, state-owned monopolist TELE Greenland announced that it had contracted Nokia Siemens Networks to roll out a LTE network in the capital Nuuk by the end of 2013, using the 800 MHz band.

7.3.14 Guatemala

Guatemala has always been something of an anomaly in terms of spectrum policy. As part of a 1996 telecoms law reform, spectrum bands were made available for any relevant purpose and an auction would only be held if competing claims were registered. The regulator would determine the parameters that governed interference but would not interfere in the choice of services to be provided. Only the spectrum bands needed for broadcasting would be excluded. As a result, 140 MHz of spectrum was brought into use for mobile networks and, after a merger took place, three powerful incumbents—Telefónica, América Móvil and Millicom—became established providers of 3G at an early stage for Latin America.

This philosophy has continued in the context of 4G. In August 2011, the regulator announced that seven bands could be used for 4G—450 MHz, 800 MHz, 900 MHz, 1,700–1,800 MHz, 1,700–2,100 MHz, 1,800–1,900 MHz and 2,000 MHz. It would be up to the operators themselves to arrange for the provision of 4G.

7.3.15 Haiti

Digicel has ostensibly launched but in practice is marketing HSPA+ as '4G'. In July 2013, the regulator stated that there would be an auction of spectrum in the 1,900 MHz band that had previously constituted a licence held by Haitel (now in receivership). The licence would be allocated to a new entrant with a view to rolling out a 3G/4G network.

7.3.16 Jamaica

In March 2013, the government announced initial plans for a single round sealed bid auction of spectrum in the 700 MHz band which, unusually, was unoccupied. Two 15-year licences would be offered with a population coverage requirement of 50 % within 18 months and 90 % within 4 years applying to incumbents and 30 % within 24 months and 90 % within 5 years applying to new entrants. Licences would be revoked if incumbents had not launched within 18 months of receiving a licence (24 months in the case of new entrants). With two well-established incumbents in LIME (Cable and Wireless) and Digicel, the expressed hope that a licence would be won by a new entrant appeared to be unrealistic (TeleGeography 2013m), especially given that the third incumbent, América Móvil, had recently sold its network to Digicel.

Altogether, 108 MHz is available. Of this, two blocks of 12 MHz paired was initially to comprise the core licences (704–715 MHz paired with 734–745—Band 17—at a reserve price of $45 million and 746–757 MHz paired with 76–787 MHz—Band 13—at a reserve price of $40 million) with the rest available in 6 MHz blocks suitable for paired or unpaired use, with no single operator allowed to exceed 12 MHz paired in the auction or 80 MHz of spectrum across all bands. However, 758–769 MHz paired with 788–799 MHz (Band 14) is not now going to be sold unless the degree of demand so warrants (TeleGeography 2013v). If a new entrant wins a core licence it will be given the bonus of a 15-year licence comprising 5 MHz paired in the 1,800 MHz or 1,900 MHz bands largely free of charge for 2 years. However, by May 2013 it had become clear that there would be no bidders at the reserve prices specified and hence the auction was postponed.

7.3.17 Mexico

The situation is unclear. In June 2010, 15 MHz in the 1,870–1,910 MHz band paired with 15 MHz in the 1,950–1,990 MHz band was offered in eight of the nine regions (Licitación 20). It would appear to have been sold for $232 million. Licitación 21 in July involved 1,710–1,755 MHz paired with 2,210–2,155 MHz in all nine regions. However, two nationwide licences each provided 15 MHz paired while three blocks of 5 MHz paired were also offered in each region. 5.248 billion pesos ($420 million) was raised in total, but due to restrictive bidding conditions one nationwide licence was bought by a Televisa/Nextel joint venture for the minimum price of 180 million pesos and the other was left unsold. América Móvil acquired most of the other spectrum for 3.794 billion pesos which it was expected to use for the introduction of LTE—tests commenced in August 2011 with a view to a launch in 2012 which allegedly took place in April without the spectrum band being specified. A subsequent report claimed that the commercial launch took place in November in an initial nine cities providing a maximum downlink of 20 Mbps, with a further 17 cities pencilled in for 2013Q1 (Telecom.paper 2012d)—it is alleged that it also intends to use WiMAX. Telefónica (Pegaso) was the other major buyer, paying a total of 1.274 billion pesos—see Landa (2010) for an analysis of the auction.

Subsequent upon the auction, some dismay was expressed about the failure of a new entrant to gain a licence, which was attributed to such factors as the dominance of América Móvil, high inter-connection fees and caps on foreign investment.

In June 2012, Telefónica announced that it was about to sign a 5-year agreement with Iusacell to share network infrastructure, including cell-sites, and that they would jointly begin testing LTE in a number of cities towards the end of 2012. Subsequently, it announced plans to launch in November in an initial six cities, but added that coverage would not be extended much during 2013.

Meanwhile MVS, the owner of a licence providing 190 MHz in the 2.6 GHz band, had been told by the regulator in February 2010 that its failure to introduce services would result in the non-renewal of its licence to provide WiMAX with the spectrum being reallocated for LTE. Eventually, in April 2011, a consortium led by MVS and including Clearwire, Intel and Alestra, put forward a proposal to launch a TD-LTE network branded 'Mobile Broadband for Everyone' which would act as a wholesaler for all-comers (Gabriel 2011). However, uncertainty about the future prospects for the licences caused talks to collapse, and in August 2012, the government reiterated its determination to recover all 2.6 GHz licences from the existing 11 licensees when they expired—some not before 2018—and put them up for auction (Harrup 2012). However, in April 2013, the Supreme Court ruled in favour of MVS retaining its spectrum.

The intention is for digital dividend spectrum in the 700 MHz band to become available by 2015. In an interesting development, the government decided in September 2012 to use the band plan being standardised across most of the Asia-Pacific region in order to maximise the potential for worldwide roaming, whereas on previous occasions it had adopted band plans in use in the USA in order to expedite cross-border roaming. It was claimed that the APT plan could support seven operators compared to four for the US model and that ten times as many subscribers could be reached. In essence, the APT plan requires that the available 108 MHz are divided into two 45 MHz blocks with a guard band above and below. How many other Latin American countries will follow suit is yet to be determined, but it is likely to prove more popular than the US plan as it can accommodate more networks which are quicker to construct, costs less to roll out and is compatible with use of the band in Asia. As of February 2013, it has already been adopted in Chile, Colombia and Ecuador (see above).

7.3.18 Nicaragua

In September 2012, the government announced that it intended to issue two new licences in order to provide competition for the two incumbents. The licences in the 1,785–1,805 MHz band included the provision of LTE and imposed conditions related to rural coverage. Seven expressions of interest were lodged. In November 2012, a licence was issued to Beijing-based R&D firm Xinwei Telecom Enterprise Group, allegedly at a price reduced heavily from the original $90 million to $20 million because of the licensee's links to the son of the President of Nicaragua. The technology of choice is allegedly the proprietary multi-carrier wireless in the local loop (McWiLL) S-CDMA (TeleGeography 2013c). It is also possible that two providers of WiMAX—a subsidiary of Russia's Scartel branded as Yota Nicaragua and IBW, an ISP—will switch to TD-LTE.

7.3.19 Paraguay

The regulator invited applications for the following spectrum: Block C (1,730–1,740 MHz paired with 2,130–2,140 MHz); Block D (1,740–1,750 MHz paired with 2,140–2,150 MHz); Block E (1,750–1,760 MHz paired with 2,150–2,160 MHz); and Block F (1,760–1,770 MHz paired with 2,160–2,170 MHz). A further Block (1,710–1,730 MHz paired with 2,110–2,130 MHz) was reserved for state-owned Copaco (Vox). Auction participants needed to have a market capitalisation of $10 million and needed to bid at least $600,000 per block with an overall cap of 40 MHz per bidder (TeleGeography 2012g).

Copaco expressed its intention to launch during Xmas 2012 in the capital Asunción and neighbouring Central department, but did not achieve this until February 2013 (Tomás 2013). Personal also launched in the same month but details are sketchy. Both appeared to provide a maximum downlink of 60 Mbps.

7.3.20 Peru

In November 2011, both Telefónica and América Móvil announced the launch of so-called 4G'. In both cases this was acknowledged to be HSPA+ (RPP 2011; Goga 2011). An auction for additional spectrum in the 900 MHz band is planned (but has been postponed). So far there is no evidence of any intention to roll out LTE in this or any other spectrum band, but this is likely to change now that there has been a successful auction in July 2013 for two 20 MHz paired blocks within the bands 1,710–1,770 MHz paired with 2,110–2,170 MHz (TeleGeography 2012v). The licences run for 20 years and the minimum bid was $63.4 million. In June 2013, Telefónica, América Móvil, Entel Chile (trading as Americatel) and Viettel Peru qualified as bidders, but only Telefónica (winning the A block for $152.2 million) and Entel (winning the B block for $105.5 million) were successful (TeleGeography 2013ae).

In a move that ran contrary to experience almost everywhere else in the world, the regulator ruled in November 2012 that it would be illegal for an operator to market its services as '4G' or 'fourth generation' when its downlink was effectively the same as it provided for 3G services (TeleGeography 2012aj).

7.3.21 Puerto Rico

In March 2008, América Móvil was awarded a licence (which it set aside for LTE use) in the 700 MHz band as part of the process taking place in mainland USA, with a requirement of 35 % population coverage within 4 years—hence the need to launch before March 2012. AT&T, which received a similar licence in October 2007, set out to follow suit using the 700 MHz and AWS bands, while Open Mobile and Sprint Nextel also intended to use their cdma2000 spectrum (TeleGeography 2011b). There was competition among the four to be the first to launch in Latin America, with AT&T winning the race in November 2011. Both América Móvil and Open Mobile announced that they would launch in San Juan in December, using LTE in the 700 MHz band, with Open Mobile's launch appearing to have taken place only in April 2012. América Móvil subsequently claimed that its initial launch had been 'soft', primarily because of a paucity of handsets, and that a commercial launch had taken place in November 2012 in 19 of the island's 78 municipalities with nation-wide coverage to follow in 2013. Even then, however, the maximum available downlink was a meagre 30 Mbps. The use of this band maximises opportunities to roam onto mainland USA.

In April 2012, Sprint Nextel announced that it would hopefully be launching its Network Vision/LTE network in San Juan in August, with existing cdma2000 1xEV-DO provision in the 1,900 MHz band integrated with LTE technology. It also proposed to use its unused legacy iDEN spectrum in the 800 MHz band to extend its service (TeleGeography 2012h). The launch took place in December. For its part, T-Mobile US did not launch until July 2013 with coverage of half the island's population, but claimed at the time that its network was faster than that of AT&T (TeleGeography 2013ad).

7.3.22 Trinidad and Tobago

In November 2012, TSTT, using the mobile brand Bmobile, launched 'Next. It's 4G, but not just 4G'. In fact, 'but not really 4G' would have been a more apt description given that it was, in fact, HSPA+ (TeleGeography 2012af).

7.3.23 Turks and Caicos Islands

In July 2012, Islandcom launched HSPA+. In November, Digicel announced that it would be following suit, but marketing its service as '4G' as it was doing elsewhere in the Caribbean. In February 2013, after 6 months of bidding, Digicel became the only operator to be awarded spectrum in the Lower B and C blocks of the 700 MHz band after more than meeting the minimum deployment conditions set by the regulator which included a launch within 18 months and 98 % population coverage within 36 months. Islandcom was awarded spectrum in the Upper C band.

7.3.24 Uruguay

In September 2011, state-owned Antel was obliged to return the 904.5–909.5 MHz and 949.5–954.5 MHz bands to the regulator. These were auctioned in March 2013 by way of 20-year licences together with 910–915 MHz paired with 955–960 MHz, 1,855–1,885 MHz paired with 1,935–1,965 MHz (6 blocks of 5 MHz paired) and 1,725–1,770 MHz paired with 2,125–2,170 MHz (9 blocks of 5 MHz paired). The auction resulted in Telefónica acquiring 40 MHz in the 1,900 MHz band for $32 million while América Móvil acquired 20 MHz in the 1,900 MHz band (1,855–1,860 MHz paired with 1,935–1,950 MHz and 1,890–1,895 MHz paired with 1,970–1,975 MHz) for $15.2 million and 20 MHz of AWS spectrum (1,745–1,755 MHz paired with 2,145–2,155 MHz) for a further $15.2 million. Antel did not participate but was awarded 10 MHz of 900 MHz spectrum together with 40 MHz of AWS spectrum (1,725–1,745 MHz paired with 2,125–2,145 MHz) for $38 million (TeleGeography 2013k). All 1,900 MHz/AWS prices were very close to the minimum of $7.5 million per 5 MHz paired. The 15 MHz paired of AWS spectrum left unsold reflected the fact that this block did not overlap with the use of the band in the USA nor was it in use elsewhere.

Meanwhile, Antel had launched over its existing AWS spectrum (1,710–1,725 MHz paired with 2,110–2,125 MHz) in December 2011. Hence, unlike in, say, Ecuador, the decision to favour the state-owned operator had nothing obvious to do with the need to bolster competition.

7.3.25 US Virgin Isles

Sprint Nextel has promised to launch using its PCS spectrum, but the only recorded launch so far is that by AT&T in July 2013 on the islands of St Croix and St Thomas

(TeleGeography 2013ab). In line with its questionable policy in mainland USA, AT&T claims that when subscribers are out of contact with the LTE network they will still be able to connect at '4G speed' to the HSPA+ network.

7.3.26 Venezuela

In May 2012, the regulator announced that it would be auctioning three nationwide licences in the following bands: Block D-D comprising 1,850–1,860 MHz paired with 1,940–1,950 MHz; Block E-E comprising 1,865–1,875 MHz paired with 1,955–1,965 MHz; and Block F-F comprising 1,715–1,730 MHz paired with 1,810–1,825 MHz. Each MHz would cost at least VEF803,000 ($187,000) annually with the expectation that the licences would be won by the three incumbents, Telefónica, Móvilnet and Digitel—as indeed proved to be the case with Digitel winning Block F-F. The spectrum would be used initially to enhance HSPA+ services, but it remained unclear whether it would also be used for LTE. For its part, Digitel expressed its desire to be awarded additional spectrum in the 1,800 MHz band for LTE and the regulator is under pressure to release spectrum in the 700 MHz band.

Telefónica, which had initially claimed that it wanted to launch LTE before the end of 2012, was duly awarded a 10 MHz paired licence in September 2012. However, it then decided to postpone any decision about a launch of LTE until 2013 pending clarification of when further licences would be issued (TeleGeography 2012y).

Móvilmax has a nationwide licence covering 48 MHz in the 2.6 GHz band, and intends to convert from WiMAX to LTE during 2013.

References

Barton, J. (2012). *New router extends local 4G/LTE coverage in Kyrgyzstan*. Accessed March 9, 2012, from http://www.developingtelecoms.com

Businessmega. (2012). *Emtel launched the 4G mobile*. Accessed May 30, 2012, from http://www.business.mega.mu

Cellular-news. (2012a). *Bahrain plans LTE spectrum sale later this year*. Accessed March 21, 2012, from http://www.cellular-news.com

Cellular-news. (2012b). *Omantel launches LTE services in Oman*. Accessed July 17, 2012, from http://www.cellular-news.com

Cellular-news. (2012c). *Bahamas regulator issues 700MHz radio spectrum licenses*. Accessed September 10, 2012, from http://www.cellular-news.com

Cellular-news. (2012d). *Vodacom launches LTE services in South Africa*. Accessed October 10, 2012, from http://www.cellular-news.com

Cellular-news. (2012e). *South Africa's Cell C planning LTE launch by end of 2012*. Accessed November 6, 2012, from http://www.cellular-news.com

Cellular-news. (2012f). *Zain launches LTE services in Kuwait*. Accessed November 25, 2012, from http://www.cellular-news.com

Cellular-news. (2013a). *Altel-Lucent wins Iraqi LTE deployment contract*. Accessed February 27, 2013, from http://www.cellular-news.com

Cellular-news. (2013b). *KT to provide LTE technology to Rwanda*. Accessed March 11, 2013, from http://www.cellular-news.com

Cellular-news. (2013c). *Ooredoo to launch LTE network next month*. Accessed March 15, 2013, from http://www.cellular-news.com

Cellular-news. (2013d). *Mozambique to auction 800MHz band*. Accessed March 18, 2013, from http://www.cellular-news.com

Cellular-news. (2013e). *Further delays expected in Israeli 4G spectrum auction*. Accessed May 13, 2013, from http://www.cellular-news.com

Cellular-news. (2013f). *Zimbabwe to make infrastructure sharing mandatory*. Accessed May 20, 2013, from http://www.cellular-news.com

Gabriel, C. (2011). *Intel and Clearwire back Mexican LTE venture*. Accessed April 9, 2011, from http://www.rethink-wireless.com

Globaltelecomsbusiness. (2012). *Colombia raises $80m in spectrum auction*. Accessed July 2, 2012, from http://www.globaltelecomsbusiness.com

Goga, A. (2011). *Movistar lanza su Red 4G (HSPA+) en Perú*. Accessed November 18, 2011, from http://www.arturogoga.com

GrupoConvergencia. (2012). *UNE-EPM got 50 MHz on the 2.5 GHz band and plans to develop LTE*. Accessed July 2, 2012, from http://www.convergencia.com

Harrup, A. (2012). *Mexican govt to recover, resell high-frequency spectrum*. Accessed August 13, 2012, from http://www.telecoms.com

Landa, R. (2010). *Spectrum auction tragedies. The case of the Mexico spectrum auction for AWS services*. Accessed September 15, 2010, from http://www.ssrn.com/abstract=1667950

Lennighan, M. (2011). *Three Saudi mobile operators claim LTE launches*. Accessed September 15, 2011, from http://www.totaltele.com

Lennighan, M. (2012). *Brazil to drive move to LTE in Latin America*. Accessed February 29, 2012, from http://www.totaltele.com

LteWorld. (2012). *Viva launches 4G LTE services in Kuwait*. Accessed October 9, 2012, from http://www.lteworld.org

MinTIC. (2013). *Gobierno adjudica licencias de 4G*. Accessed June 26, 2013, from http://www.mintic.gov.co

Prescott, R. (2012). *Bolivia's Entel launches LTE in 700 MHz band*. Accessed December 22, 2008, from http://www.rcrwireless.com

Rayal, F. (2012). *Analysis of Brazil's 2.5 GHz spectrum auction results*. Accessed June 19, 2012, from http://www.frankrayal.com

RPP. (2011). *Claro lanza tecnologia 4G en el Perú*. Accessed November 15, 2011, from http://www.rpp.com.pe

Telecom.paper. (2012a). *Azercell launches LTE in Baku*. Accessed May 24, 2012, from http://www.telecompaper.com

Telecom.paper. (2012b). *Colombia to award 4G spectrum in September*. Accessed May 28, 2012, from http://www.telecompaper.com

Telecom.paper. (2012c). *Kyrgyzstan allocates LTE, Wimax frequencies to 12 operators*. Accessed October 10, 2012, from http://www.telecompaper.com

Telecom.paper. (2012d). *Telcel launches LTE network in 9 Mexican cities*. Accessed November 7, 2012, from http://www.telecompaper.com

Telecom.paper. (2012e). *Orange Moldova launches LTE services*. Accessed November 21, 2012, from http://www.telecompaper.com

Telecom.paper. (2012f). *Jordan seeks new operators in spectrum tender*. Accessed December 7, 2012, from http://www.telecompaper.com

Telecom.paper. (2012g). *CNT wins 70 MHz spectrum for LTE services*. Accessed December 14, 2012, from http://www.telecompaper.com

Telecom.paper. (2012h). *Claro Brasil starts 4G services*. Accessed December 14, 2012, from http://www.telecompaper.com

Telecom.paper. (2012i). *Kazakhtelecom launches LTE in Astana, Almaty.* Accessed December 27, 2012, from http://www.telecompaper.com

Telecom.paper. (2013a). *Israeli govt to allocate 4G frequencies in 2013.* Accessed January 9, 2013, from http://www.telecompaper.com

Telecom.paper. (2013b). *Tricom launches LTE network in Dominican Republic.* Accessed March 27, 2013, from http://www.telecompaper.com

Telecom.paper. (2013c). *Algeria to close Djezzy purchase before 3G licensing.* Accessed April 8, 2013, from http://www.telecompaper.com

Telecom.paper. (2013d). *Claro Brasil launches 4G service in 11 cities.* Accessed April 18, 2013, from http://www.telecompaper.com

Telecom.paper. (2013e). *Claro launches first LTE network in Chile.* Accessed June 27, 2013, from http://www.telecompaper.com

TeleGeography. (2010). *Fourth LTE network goes live: MTS launches 4G in Tashkent.* Accessed July 29, 2010, from http://www.telegeography.com

TeleGeography. (2011a). *Airtel, Yu face foreign ownership LTE licence freeze.* Accessed September 5, 2011, from http://www.telegeography.com

TeleGeography. (2011b). *Race is on for LTE in Puerto Rico.* Accessed October 4, 2011, from http://www.telegeography.com

TeleGeography. (2011c). *WBS announces LTE rollout plans: Launch expected by mid-2012.* Accessed October 12, 2011, from http://www.telegeography.com

TeleGeography. (2012a). *DISH LTE plan moves forward in Canada, not in US.* Accessed February 9, 2012, from http://www.telegeography.com

TeleGeography. (2012b). *TRA invites expressions of interest in unallocated 1900MHz-2170MHz.* Accessed March 2, 2012, from http://www.telegeography.com

TeleGeography. (2012c). *ICASA bows to operator pressure; postpones 800MHz, 2.6GHz spectrum auction until further notice.* Accessed March 7, 2012, from http://www.telegeography.com

TeleGeography. (2012d). *Spectrum policy addresses 4G needs, but no timetable.* Accessed March 21, 2012, from http://www.telegeography.com

TeleGeography. (2012e). *NCC looks to free up 2.5 GHz spectrum.* Accessed March 20, 2012, from http://www.telegeography.com

TeleGeography. (2012f). *BTC upgrades Abaco to '4G'.* Accessed April 16, 2012, from http://www.telegeography.com

TeleGeography. (2012g). *Conatel demands second frequency consultation regarding 1700MHz and 2100MHz bands.* Accessed April 17, 2012, from http://www.telegeography.com

TeleGeography. (2012h). *Sprint Puerto Rican LTE launch anticipated in third quarter.* Accessed April 19, 2012, from http://www.telegeography.com

TeleGeography. (2012i). *No day of rest as Yota Bel sets Sunday launch date for LTE.* Accessed May 5, 2012, from http://www.telegeography.com

TeleGeography. (2012j). *Subtel receives three proposals for LTE, Entel comes out top.* Accessed May 14, 2012, from http://www.telegeography.com

TeleGeography. (2012k). *LTE arrives in Namibia: MTC launches 4G in capital.* Accessed May 18, 2012, from http://www.telegeography.com

TeleGeography. (2012l). *STC adds 4G coverage across 1,500 sites.* Accessed May 30, 2012, from http://www.telegeography.com

TeleGeography. (2012m). *Minister sets out LTE timeline; launch set for 2013, nationwide coverage by 2018.* Accessed May 30, 2012, from http://www.telegeography.com

TeleGeography. (2012n). *Smile launches 4G network in Dar es Salaam.* Accessed June 8, 2012, from http://www.telegeography.com

TeleGeography. (2012o). *Du eyeing launch of LTE next month.* Accessed June 11, 2012, from http://www.telegeography.com

TeleGeography. (2012p). *Cellcom Liberia launches '4G' network.* Accessed June 25, 2012, from http://www.telegeography.com

TeleGeography. (2012q). *LIME claims '4G' HSPA+ launch in BVI*. Accessed July 2, 2012, from http://www.telegeography.com

TeleGeography. (2012r). *Anrceti opens 4G licensing public consultation*. Accessed July 30, 2012, from http://www.telegeography.com

TeleGeography. (2012s). *Subtel allocates 4G spectrum*. Accessed July 31, 2012, from http://www.telegeography.com

TeleGeography. (2012t). *MTN's South African unit to launch LTE, report says*. Accessed August 13, 2012, from http://www.telegeography.com

TeleGeography. (2012u). *Antigua's govt pushes ahead with digital switchover; hints at LTE licensing in 1Q13*. Accessed August 14, 2012, from http://www.telegeography.com

TeleGeography. (2012v). *4G sale set for 1H 2013*. Accessed August 17, 2012, from http://www.telegeography.com

TeleGeography. (2012w). *URCA grants BTC 4G licence*. Accessed September 3, 2012, from http://www.telegeography.com

TeleGeography. (2012x). *Confusion over Orange Dominicana's 4G claims*. Accessed September 6, 2012, from http://www.telegeography.com

TeleGeography. (2012y). *Movistar wins extra 1900MHz spectrum, ups investment to USD535m, delays LTE to 2014*. Accessed September 24, 2012, from http://www.telegeography.com

TeleGeography. (2012z). *Kuwait licenses 4G services in 1800MHz band*. Accessed October 8, 2012, from http://www.telegeography.com

TeleGeography. (2013a). *Bolivia spectrum sale expected in 1Q 2013*. Accessed January 7, 2013, from http://www.telegeography.com

TeleGeography. (2013b). *Indotel lifts ban on Orange's 4G service*. Accessed January 15, 2013, from http://www.telegeography.com

TeleGeography. (2013c). *Claro offers new licensee Xinwei a warm welcome as plot thickens round Chinese firm*. Accessed January 18, 2013, from http://www.telegeography.com

TeleGeography. (2013d). *Qtel focused on network improvements, 4G launch this year*. Accessed January 17, 2013, from http://www.telegeography.com

TeleGeography. (2013e). *TRA issues 2600MHz, 2100 MHz, 1800 MHz, 900 MHz invitation*. Accessed January 28, 2013, from http://www.telegeography.com

TeleGeography. (2013f). *EETT consults on 3.4GHz-3.8GHz allocation, usage*. Accessed February 11, 2013, from http://www.telegeography.com

TeleGeography. (2013g). *Batelco launches commercial LTE-1800 services*. Accessed March 1, 2013, from http://www.telegeography.com

TeleGeography. (2013h). *Nawras introduces LTE services*. Accessed March 5, 2013, from http://www.telegeography.com

TeleGeography. (2013i). *TIM Brasil board gives thumbs up to network sharing plan with Oi SA*. Accessed March 6, 2013, from http://www.telegeography.com

TeleGeography. (2013j). *Umniah tests LTE network, Orange 'not part of the game'*. Accessed March 11, 2013, from http://www.telegeography.com

TeleGeography. (2013k). *Regulator assigns 900MHz, 1900MHz, 1700MHz/2100MHz spectrum*. Accessed March 15, 2013, from http://www.telegeography.com

TeleGeography. (2013l). *Menatelecom not allowed to bid in 4G auction*. Accessed March 18, 2013, from http://www.telegeography.com

TeleGeography. (2013m). *700MHz spectrum up for grabs; MTRs may fall to JMD2*. Accessed March 27, 2013, from http://www.telegeography.com

TeleGeography. (2013n). *Turkish government inks 4G deal*. Accessed April 9, 2013, form http://www.telegeography.com

TeleGeography. (2013o). *Zain Bahrain launches 4G network*. Accessed April 11, 2013, from http://www.telegeography.com

TeleGeography. (2013p). *2Connect attempts to sue TRA over LTE auction exclusion*. Accessed April 11, 2013, form http://www.telegeography.com

TeleGeography. (2013q). *Regulator announces results of consultation on 2500MHz-2690MHz licence auction.* Accessed April 23, 2013, from http://www.telegeography.com

TeleGeography. (2013r). *Belarusian minister confirms plans for LTE-A tests in 2013.* Accessed April 25, 2013, from http://www.telegeography.com

TeleGeography. (2013s). *KT LTE network generates KZT100m, claims 30,000 4G subs.* Accessed April 26, 2013, from http://www.telegeography.com

TeleGeography. (2013t). *Oi SA inaugurates Rio 4G launch.* Accessed April 26, 2013, from http://www.telegeography.com

TeleGeography. (2013u). *MTN activates Uganda's first LTE network.* Accessed April 29, 2013, from http://www.telegeography.com

TeleGeography. (2013v). *SMA sets out plans for 700 MHz auction, sweetens deal for newcomers.* Accessed April 29, 2013, from http://www.telegeography.com

TeleGeography. (2013w). *Alfa launches commercial LTE in Beirut.* Accessed May 20, 2013, from http://www.telegeography.com

TeleGeography. (2013x). *Touch launches commercial LTE.* Accessed May 23, 2013, from http://www.telegeography.com

TeleGeography. (2013y). *Smile launches 800MHz LTE.* Accessed June 11, 2013, from http://www.telegeography.com

TeleGeography. (2013z). *Zambia's Massnet seeking investor to enable LTE deployment.* Accessed June 21, 2013, from http://www.telegeography.com

TeleGeography. (2012aa). *APEK approves LTE 700MHz tests.* Accessed October 15, 2012, from http://www.telegeography.com

TeleGeography. (2012ab). *Rogers confirms 2600MHz LTE availability across its 2100MHz footprint.* Accessed November 1, 2012, from http://www.telegeography.com

TeleGeography. (2012ac). *Ukrtelecom issued nationwide 5.3GHz wireless broadband licence.* Accessed November 9, 2012, from http://www.telegeography.com

TeleGeography. (2012ad). *Neotel announces 70Mbps LTE trial.* Accessed November 12, 2012, from http://www.telegeography.com

TeleGeography. (2012ae). *Colombia sets the bar for LTE providers.* Accessed November 14, 2012, from http://www.telegeography.com

TeleGeography. (2012af). *This is not just any '4G', this is TSTT '4G'.* Accessed November 14, 2012, from http://www.telegeography.com

TeleGeography. (2012ag). *Smile Telecoms selects Alca-Lu to roll out 4G networks in Tanzania, Uganda and DRC.* Accessed November 16, 2012, from http://www.telegeography.com

TeleGeography. (2012ah). *TRA speeds up release of 2.5GHz-2.6GHz bands.* Accessed November 20, 2012, from http://www.telegeography.com

TeleGeography. (2012ai). *Unknown company awarded 1980MHz-2000MHz licence.* Accessed November 23, 2012, from http://www.telegeography.com

TeleGeography. (2012aj). *Peru bans use of '4G' in marketing.* Accessed November 28, 2012, from http://www.telegeography.com

TeleGeography. (2012ak). *MTN LTE network now up and running in Pretoria, Durban, Jo'burg.* Accessed December 5, 2012, from http://www.telegeography.com

TeleGeography. (2012al). *Menatelecom investing in nationwide LTE rollout.* Accessed December 6, 2012, from http://www.telegeography.com

TeleGeography. (2013aa). *Spectrum transfer framework: Looks good for Verizon, bad for Rogers.* Accessed July 1, 2013, from http://www.telegeography.com

TeleGeography. (2013ab). *USVI receives LTE courtesy of AT&T; US 4G coverage reaches 326 cities.* Accessed July 3, 2013, from http://www.telegeography.com

TeleGeography. (2013ac). *Econet to roll out 4G LTE in Victoria Falls.* Accessed July 9, 2013, from http://www.telegeography.com

TeleGeography. (2013ad). *T-Mobile launches LTE in Puerto Rico.* Accessed July 18, 2013, from http://www.telegeography.com

TeleGeography. (2013ae). *Telefonica, Entel win 4G spectrum tender in Peru.* Accessed July 23, 2013, from http://www.telegeography.com

TeleGeography. (2013af). *Fed-up Telus takes government to court over spectrum transfer rules.* Accessed July 30, 2013, from http://www.telegeography.com

The Media Guru. (2012). *Orange Mauritius launches 4G LTE!* Accessed June 29, 2012, from http://www.themediaguru.blogspot.com

Tomás, J. (2013). *State telco Copaco investing in Paraguay LTE developments.* Accessed February 21, 2013, from http://www.totaltele.com

Wireless Federation. (2011). *Argentina to auction mobile spectrum licenses in H2.* Accessed May 12, 2011, from http://www.wimaxforum.org/about

Conclusions from Case Studies

8.1 LTE Subscriber Growth

Where possible in the case studies, some indication of subscriber numbers has been given. As of end-2011, these were only significant in the USA, Japan and South Korea, with the USA accounting for more than half of the total and, in turn, Verizon Wireless accounting for almost all of the US total. However, bearing in mind that the number of subscribers at the end of 2010 was estimated at between 50,000 and 100,000, the achievement of between 8.5 million and 9 million by the end of 2011 (TeleGeography 2012b) indicated an exponential increase that was likely to be repeated during 2012.[1] According to 4G-Portal, 38 million connections existed at the end of 2012, with that number rising to 100 million in May 2013. Of these, 57 million were recorded in North America (Cellular-news 2013b). According to the World Cellular Information Service [as reported in Mobile Communications International June (2013: p. 5)], the total as of March 2013 was 88.5 million (up from 17.8 million 1 year earlier) with 48.9 million attributed to USA/Canada and 35.7 million to Asia Pacific.

The 46.9 million attributed solely to the USA by the WCIS represented a penetration rate of 14.8 %, somewhat above the 11 % penetration achieved in Japan (equivalent to 13.8 million subscribers). However, by far and away the greatest penetration rate was measured in South Korea at 40.2 % (equivalent to 19.6 million subscribers). Hence, these three countries accounted for 80.4 million of the total or 90.8 %.

According to the WCIS, the top ten operators in terms of LTE subscribers in March 2013 comprised four in the USA—Verizon Wireless (26.3 million), AT&T (10.5 million), Sprint Nextel (6.6 million) and MetroPCS (3.5 million)—three in Korea—SK Telecom (9.3 million), LG Uplus (5.2 million) and Korea Telecom (5.1

---

[1] For those with deep pockets (which excludes the authors), a number of consultancies produce forecasts of LTE subscriber growth and related matters. For example, a consultancy heavily involved with 4G is Eogogicsinc—see Eogogicsinc (2012a, b).

P. Curwen and J. Whalley, *Fourth Generation Mobile Communication*,
Management for Professionals, DOI 10.1007/978-3-319-02210-9_8,
© Springer International Publishing Switzerland 2013

million)—two in Japan—NTT DoCoMo (11.6 million) and Softbank (1.3 million)—and one in Australia—Telstra (2 million). The absence of European operators is worthy of note. The only operators that (so far) publish subscriber numbers with any regularity in the public domain appear to be SK Telecom, NTT DoCoMo and Verizon Wireless with Korea Telecom and MetroPCS also helpful bearing in mind their relatively recent launches. The most useful alternative way to summarise the data is in terms of the number of months it has taken these companies to reach certain milestones. It would appear that operators in South Korea were able to achieve one million subscribers within 6 months, as was Verizon Wireless, whereas the norm elsewhere is roughly 1 year by which time the South Korean operators had reached roughly four million. It appears that the most successful operators that have reached the ten million mile stone at the time of writing—Verizon Wireless, SK Telecom and DoCoMo—have taken between 17 and 26 months to achieve this, a highly commendable performance.

Such evidence as exists for Europe is not particularly favourable. Obviously, there are many possible explanations for different penetration rates in different countries such as subscription and device prices, the price and transfer speed of existing networks as well as cultural differences, so it is still too early to draw any general conclusions.

8.2 Cost of Licences

As is well-known, some of the early attempts at issuing licences for 3G involved large sums of revenue for government coffers. What tends to stick in everyone's minds are the $45.9 billion generated in Germany (equivalent to $560 per head of population or $93 per pop per licence) and the $34.2 billion generated in the UK (equivalent to $585 per head of population or $117 per pop per licence), or possibly the $13.9 billion raised in the AWS-1 auction in the USA. However, these were very untypical and the great majority of licence sales raised at most several billion dollars but often in the hundreds of millions.

The most notable feature of the 4G auctions so far has been an avowed desire not to treat them as revenue-raising exercises but rather to seek wider societal gains, although this laudable aim appears to have been subverted somewhat at a time of austerity. Whereas some sources are happy to publish data purporting to illustrate the cost per megahertz or some such indicator as though it was a highly meaningful datum, and some examples are available in Table 8.1, the data in the case studies cited previously are very difficult to interpret and compare. The reasons for this include the following:

1. Widely different amounts of spectrum have been involved in individual cases—it must be borne in mind that some spectrum, especially digital dividend spectrum, is more valuable because it is better suited for 4G signals.
2. Some auctions have involved single bands while others have involved multiple bands.
3. The licences have generally been issued on a technology-neutral basis.

Table 8.1 Illustrative results of sales of spectrum with potential 4G usage

Country	Date	Band (MHz)	MHz (sold)	Revenue (million)	Revenue $ (million)	$ per MHz (per pop)[a]
Austria	Sep 2010	2,600	190	€40	52	0.03
Belgium	Nov 2011	2,600	155[b]	€77.8	104	0.05
Brazil	Jun 2012	2,600	120	R2,625	1,270	0.05
Colombia	Jun 2010	2,600	50	P80,000	41	0.19
Colombia	Aug 2011	1,900	25	$80	80	0.07
Colombia	June 2013	Multiple	190	COP770,000	399	0.04[c]
Croatia	Oct 2012	800	40[d]	€40	52	0.30
Denmark	May 2010	2,600	200[e]	DKK1,010	168	0.15
Denmark	Sep 2010	900/1,800	30	DKK12	2.1	0.01
Denmark	Jun 2012	800	60	DKK742	125	0.38
Finland	Nov 2009	2,600	190	€3.8	5.7	neg.
France	Sep 2011	2,600	140	€936	1,280	0.14
France	Dec 2011	800	60	€2,639	4,480	0.95
Germany	May 2010	Multiple	359	€4,380	5,960	f
Greece	Nov 2011	900/1,800	70/40	€381	515	0.41
Hong Kong	Jan 2009	2,300/2,600[g]	90	HK$1,536	198	0.31[h]
Hong Kong	Apr 2011	850/2,000[i]	20	HK$1,925	244	1.78
Hong Kong	Feb 2012	2,300	90	HK$481	61	0.10
Hong Kong	Mar 2013	2,600	50	HK$1,540	198	0.56
India	Jun 2010	2,300	60	$8,230	8,230	0.12
Italy	Sep 2011	800/1,800/2,600	240	€3,945	5,300	0.36[j]
Latvia	Jan 2012	2,600	140[k]	LVL2,341	4,390	0.01
Macedonia	Jul 2013	800/1,800	150	€31	39	0.12[l]
Mexico	Jul 2010	AWS	60	P5,248	409	0.07
Netherlands	Apr 2010	2,600	130[m]	€2.6	3.2	neg.
Netherlands	Dec 2012	Multiple	360	€3,805	4,870	n

(continued)

Table 8.1 (continued)

Country	Date	Band (MHz)	MHz (sold)	Revenue (million)	Revenue $ (million)	$ per MHz (per pop)[a]
Peru	Jul 2013	AWS	80	$258	258	0.10
Poland	Feb 2013	1,800	50	€228	301	0.16
Portugal	Oct 2011	Multiple	299[o]	€372	510	0.17
Romania	Sep 2012	Multiple	485[p]	€682	880	0.08
Singapore	May 2005	2,300[q]	50	S$2,420	1,466	neg.
Singapore	May 2005	2,600[q]	90	S$7,419	4,496	0.01
Singapore	June 2013	1,800/2,600	270	S$360	284	0.20[r]
Spain	Jul 2011	800/900/2,600	210[s]	€1,650	2,380	0.26[t]
Sweden	May 2008	2,600	190	€225	347	0.20
Sweden	Mar 2011	800	60	SEK2,054	325	0.59
Sweden	Oct 2011	1,800	70	€148	203	0.32
Switzerland	Feb 2012	Multiple	575	CHF997	1,090	0.24

neg. = negligible (less than $0.01 per MHz per pop)

[a]Other sources tend to give slightly different values for entries in this column. It is generally not possible to work out why this is the case as they tend not to provide their underlying data. It may be noted that the Euro is often used rather than the dollar

[b]15 MHz paired of the available 70 MHz of paired spectrum was not sold

[c]The 35 MHz not sold comprised 15 MHz paired in the 2.6 GHz band and 2.5 MHz paired in the 1,900 MHz band. The 45 MHz paired of AWS spectrum fetched $259.4 million, equivalent to $0.06 per MHz per pop; the 30 MHz paired in the 2.6 GHz band fetched $99.4 million, equivalent to $0.04 per MHz per pop; and the 40 MHz unpaired fetched $40.2 million, equivalent to $0.02 per MHz per pop

[d]Only two of the available three (equal-sized) licences were sold

[e]Including 15 MHz in the 2,010–2,025 MHz band

[f]The overall figure was $0.27 per MHz per pop, but this disguised big variations: 800 MHz=$0.27 per MHz per pop

[g]All 2,300 MHz spectrum and some 2.6 GHz spectrum were left unsold

[h]Average for both bands, but the 2,600 MHz band spectrum sold for a higher price per MHz per pop

[i]The two 850 MHz licences were sold but the 2 GHz licence was left unsold

[j]The overall figure was $0.36 per MHz per pop, but this disguised big variations: 800 MHz=$1.10; 1,800 MHz=$0.35; 2.6 GHz=$0.08 for FDD, $0.06 for TDD

[k]Comprising 70 MHz paired. A 50 MHz unpaired block was left unsold

[l]60 MHz in the 800 MHz band and 90 MHz in the 1,800 MHz band was sold but the licence fees related to dual-band bundles

[m]Comprising only the paired spectrum on offer as no bids were made for the unpaired spectrum

[n]The complex mix of spectrum bands, paired/unpaired spectrum and licence lengths precludes the calculation of a meaningful average

[o]Spectrum in only three of the available six bands (800 MHz, 1,800 MHz and 2.6 GHz) attracted competitive bidding, no spectrum was sold in the 450 MHz and 2,100 MHz bands and not all of the available spectrum was sold in the 900 MHz and 1,800 MHz bands

[p]Of which 110 MHz was for a 14-month period only. All blocks in the 900 MHz and 1,800 MHz bands were sold, but one block in the 800 MHz band and eight blocks in the 2.6 GHz band were left unsold

[q]Initially issued for WiMAX use

[r]150 MHz in the 1,800 MHz band sold for $0.24 per MHz per pop whereas 120 MHz in the 2.6 GHz band sold for 0.15 per MHz per pop

[s]270 MHz was on offer, of which 50 MHz of unpaired 2.6 GHz spectrum and 4.8 MHz paired of 900 MHz spectrum, together with a regional licence in Extremadura were left unsold

[t]This figure applies only to the 190 MHz of nationwide spectrum won by the three incumbents. As in Germany, 800 MHz band spectrum sold for considerably more per MHz than 2.6 GHz spectrum

Source: Compiled by authors

4. Licence terms, including their length and coverage requirements, have varied.
5. The number of potential new entrants has been variable, although they have generally been too few to drive up prices significantly.

There is a tendency to refer to the tiny amounts raised in, for example, Norway (€29 million), but in Germany, Italy and Spain and, outside the European Union, in India and Singapore, the sums raised can be counted in the low billions of dollars and it is clear that the respective governments were being careful to (more or less) maximise revenues irrespective of their avowed desire to pursue loftier ambitions.

Leaving aside Hong Kong, where spectrum prices are always unusually high irrespective of the band on offer, it is evident from Table 8.1 that spectrum below 1 GHz is typically far more valuable than that in higher bands. One obvious explanation for this relates to the number of base stations which, for example, more than double when using the 1,800 MHz band rather than the 900 MHz band for a given population coverage (Lundborg et al. 2013).

Although it is as yet too early to analyse the effect of disparities in reserve prices, it may be noted that there is some dispute about the desirability of fixing a low reserve in order to attract new entrants. It can be argued that if solely the incumbents are realistically going to enter the bidding then the sensible approach is to set a high reserve. However, that may in turn prove to be counter-productive where the incumbents are simply unwilling to pay the reserve prices and as a result either refuse to bid at all or restrict their bids to favoured blocks of spectrum. Certainly, there have been cases where both outcomes have been observed, as noted in the case studies and Table 8.1.

Some analysts believe that operators could, and should, avoid paying inflated fees for dedicated 4G spectrum. Aircom International, for example, claimed that, on the basis of fees paid up to September 2011, re-farming a 5 MHz band of 2G/3G spectrum for LTE deployment could be done at a mere half a per cent of the cost of the equivalent new spectrum suitable for LTE. According to Aircom, re-farming would suffice for LTE provision until 2014, although some additional spectrum would obviously need to be purchased to cover demands beyond that date (Ng 2011).

8.3   Vendor Strategies

There can be no doubt that 4G is hugely important for equipment vendors. 2012 proved to be the worst year for orders since 2003, with operators holding back on investment in network infrastructure as they themselves react to the uncertainties and, in many cases, falling subscriber numbers, induced by, for example, recent events in Europe (Cellular-news 2012k). During 2012Q2, for example, the only growth that occurred was in spending on LTE with the top three vendors listed as Ericsson, Alcatel-Lucent and Nokia Siemens Networks (Cellular-news 2012m).

However, as is clear from the above, vendors find themselves in something of a strategic bind because there are so many spectrum bands that can potentially be used for 4G—see Newlands (2013) for an analysis in relation to the iPhone 5. While it is true that re-farming of 2G spectrum for 3G is now fairly commonplace, this was not considered in Europe in the early days of 3G. In other regions, 2G spectrum tended only to be used for 3G where operators had spare capacity and the regulator did not have a strategy for the issuing of specific 3G licences.

Currently, the choice in most regions—the USA is something of an exception— lies between digital dividend spectrum, re-farmed 2G spectrum, 3G spectrum and 2.3–2.6 GHz band spectrum. But these are not uniform across the world as noted in Chap. 2, so vendors need to develop different multi-band devices for different continents which much reduces the potential for economies of scale and hence results in unnecessarily high prices. However, it may be noted that whereas much of a cdma2000 network has to be switched off if the network is progressing via LTE, the increasingly common practice where GSM-based networks are concerned is for a major upgrade of an existing 2G/3G network to be accompanied by in-built provision for LTE, which provides a source of overall cost reduction (Gabriel 2012b; Telecom.paper 2012d).

As things stand, it would appear that, in Europe, devices will need to provide for the digital dividend, 1,800 MHz, 3G and 2.6 GHz spectrum bands, but GSM 900 MHz will not normally be used. The real problem lies in the Asia-Pacific region since operators in different countries clearly have different needs and, unlike elsewhere, the 2.3 GHz band is a key component of 4G in India. It must be borne in mind that, even in the latter case, there is a need to roam internationally so at least one other band, presumably the 2.6 GHz band, will need to be accessible in devices.

The situation inside the USA is even more complicated since (as noted in Chap. 3) there is an issue arising from the need to roam within bands as well as across bands. Within the 700 MHz band there are four band classes. Small operators use Band Class 12 for 4G whereas AT&T uses Band Class 17 and Verizon primarily Band Class 13. Because the smaller operators will be disadvantaged if they cannot roam freely onto the networks of the major operators, the FCC is considering making this mandatory (Gabriel 2012d).

A further factor is the split between FDD and TDD. Originally, there seemed to be little incentive to design dual-mode devices as TD-LTE appeared to have poor prospects. With the switch-over from WiMAX to TD-LTE now assured in many quarters, the need for dual-mode devices has risen markedly. However, dual-mode/multiple-band devices are not going to be cost-efficient to develop, so vendors may face awkward questions as to where to prioritise their resources. In particular, there is an interesting problem in the USA where Verizon and AT&T are concentrating on FDD while Clearwire needs to utilise TD-LTE and multiple spectrum bands are in usage. It is, of course, to be expected that dual-mode issues will be resolved in the near future as technological progress waits for no man, and at present ZTE appears to be the most likely vendor to make progress on this front.

One issue, at least, which is helping to simplify matters for vendors concerns the choice of operating system for their devices. According to IDC, 722 million smartphones were shipped during 2012, an annual increase of 46 %. Of these, 136 million used Apple's iOS while 497 million used Android. BlackBerry,

Symbian and Windows Phone shared what was left, but given the increasing dominance of iOS and Android, these may or may not stay the course (Wood 2013).

A final issue concerns the decision whether to proceed as a single vendor or via a partnership of some kind. It is of interest that, in late August 2011, NEC announced that it would be collaborating with Cisco to build its LTE networks (Cellular-news 2011c).

During 2010/2011 the standard approach was to manufacture single-band dongles and dual-band handheld devices as indicated in footnote 6—according to the Global mobile Suppliers Association, almost exactly half of the 350 devices available in May 2012 were configured for the 700 MHz band in use in the USA with a further 100 configured for the 2.6 GHz band (Wood 2012). Devices capable of 2G/3G/LTE connectivity—possibly including LTE connectivity in more than one band—began to appear during 2012 (see, for example, Telecom.paper 2012a, b). It is worth noting that adding bands necessarily adds bulk to a device, so the tablet form seems better suited than the handset form to the needs of a four-or-more-band device. The first multi-band device to be certified by the Global Certification Forum (GCF) was a tablet, the Samsung Galaxy Tab 10.1 LTE (SC-01D).

The Global mobile Suppliers Association keeps track of LTE devices via a series of reports. The first of 2013 covered the position on 1 February 2013 and revealed that 666 had been launched, including 221 smartphones—the fastest-rising category—102 dongles and 53 tablets. Some of these were either country or operator specific, with two-thirds able to connect with HSPA networks—with half of these in turn able to connect with DC-HSPA+ networks. As indicated above, the entire 700 MHz band (Bands 12, 13, 14, 17) was by far the most popular followed by the 2.6 GHz band (Band 7) and 1800 MHz band (Band 3). Moderate support was shown for TD-LTE Bands 38 and 40 (Global mobile Suppliers Association 2013a).

A follow-up report in May (Global mobile Suppliers Association 2013b) revealed that 97 vendors had announced 821 devices including carrier and frequency variants, and that the number of vendors had increased by 54 % and the number of devices by 474 during the previous year. Most notably, the number of LTE-enabled smartphones had increased by 400 %. Although the largest number of FDD devices were in the combined 700 MHz bands, the 2.6 GHz, 1,800 MHz, 800 MHz (Band 20), AWS (Band 4), 2.1 GHz (Band 1) and 800/1,800/2,100 MHz bands had achieved significant numbers of devices. On the TD-LTE front, devices for 2.3 GHz (Band 38) and 2.6 GHz (Band 40) continued to constitute the majority.

Progress is increasingly controlled by the GCF—for details see http://www.globalcertificationforum.org. This body initiated LTE device certification[2] in December 2010, and by February 2012 it had certified devices for Asia, Europe and the USA, encompassing in total 15 devices from eight vendors of which five incorporated dual-band LTE. At this point, nine bands were involved as a result of

---

[2] This should not be confused with vendors' contributions to RAN standards—see Cellular-news (2012b).

the recent addition of the 1,800 MHz and US AWS bands, but progress was expected to be rapid throughout 2012 and beyond—for example, Option announced a quad-band (800/900/1,800/2,600 MHz) dongle in April 2012 (Cellular-news 2012f).

An intriguing situation in respect of the form of connectivity is coming increasingly to the fore. In principle, one would expect the great majority of handheld devices to connect via a standard 4G signal. However, the number of sources for Wi-Fi signals has grown exponentially of late such that the newest devices, particularly tablets, are increasingly offered with either Wi-Fi or combined 4G/Wi-Fi connectivity. Given that the devices providing dual connectivity are significantly more expensive, urban dwellers are opting for Wi-Fi—roughly 80 % of tablets in the USA are Wi-Fi only. By and large, the owners of tablets and lap-tops also own a smartphone with a high-speed connection, and it is possible to 'tether' this to other devices such that the latter do not need their own (expensive) data plan. Tethering effectively transforms the handset into a Wi-Fi hotspot with the high-speed connection operating as backhaul, and with most devices it is a quick process to provide the link. Needless to say, mobile operators are less than enthused about tethering, and use various methods to try to restrict it, but it is almost certainly a losing battle that is being waged (Lewis 2012).

Meanwhile component makers are also working hard to develop silicon devices with features such as Carrier Aggregation and multi-user MIMO to support LTE-Advanced. For example, Marvell launched the PXA1802 in March 2012 which integrated both TDD and FDD versions of LTE with TD-SCDMA (Telecom.paper 2012c; Gabriel 2013a). Components are categorised as Cat. 2, supporting 50 MHz (rapidly becoming obsolete); Cat. 3, supporting 100 MHz; and Cat. 4, supporting 150 MHz—see, for example, Gabriel (2012a). Recent developments include Broadcom's BCM21892 Category 4 modem (4G-Portal 2013) and Nvidia's 2.3 GHz Tegra's 4i LTE smartphone processor (Telecom.paper 2013). According to a recent report in May 2013 (Cellular-news 2013a), at that point in time four vendors were offering integrated platforms and a further four were offering Carrier Aggregation. In addition, 12 vendors were offering 23 products supporting Cat. 3 and 5 were offering 10 products supporting Cat.4.

On the antenna front, Antenova has unveiled a wideband LTE antenna reference design for smartphones which precludes the need to use switches or active tuneable components (Gabriel 2012f). In order to remain competitive in such a fast-moving market, component manufacturers such as Intel, Nvidia, Renesas and ST-Ericsson may have to engage in M&A activity. However, the extremely rapid growth in the market for LTE devices does mean that the suppliers of those components that are in short supply—displays and basebands in mid-2012—are able to take advantage in the short term (Gabriel 2012c).

A final issue for vendors relates to patents. As of mid-2012, roughly 3,000 LTE patents had been filed, double that for 3G. Under the circumstances, if all patent holders try to charge as much as possible for licensing, then royalties will potentially push up the costs of hardware to uneconomic levels. Standard practice in the mobile sector is for patent owners to strike secret bilateral deals, and with vendors

under increasing financial pressure there are reasons to expect this to continue—
Nokia, for one, intends to carry on milking its patents. However, there is pressure
for a patent pool to be formed whereby all participants sign up to a common
framework and royalty structure for everyone wishing to license a given payment.
So far, attempts to form a widespread pool have foundered, but VIA Licensing is
the latest to give it a shot (Gabriel 2012e, g). Meanwhile, the incentive to take legal
action remains (Dignan 2012).

General Conclusions

As noted above, the excitement generated by the arrival of 4G was subdued at
best. This could be accounted for by two main factors: Firstly, there was a
universal desire to avoid a repetition of what happened when 3G licences were
issued and, secondly, the development of HSPA+ meant that the initial data
transfer speeds that could realistically be provided by 4G offered only modest
improvements on what was already available. It may also be added that operators
are well aware that it is becoming more and more difficult to create sustainable
differentiation where all networks provide LTE and all operators provide essentially the same range of devices.

It is easy to lose sight of the fact that HSPA+ is still being wheeled out for the
first time in many countries, but even in those with established networks the
transfer speed and geographical coverage still have some way to go. Furthermore, technical improvements are still in progress. For example, in February
2013, Huawei and Qualcomm demonstrated the feasibility of Scalable UMTS
which supports 3G within a half-bandwidth channel (White 2013). The underlying issue here is the desire to re-farm the 900 MHz band despite the fact that
most operators have no more than 5 MHz of contiguous spectrum, which is
insufficient for LTE. By in effect doubling this bandwidth without any loss of
spectral efficiency, the vendors expect, subject to authorisation by 3GPP, to
improve the prospects for use of the popular 900 MHz band for 4G.

Operator caution initially also reflected the inevitable delays in developing
customer equipment. All of the initial launches used modems/dongles, and while
an adequate range of handheld devices was announced at CES 2011—there were
the usual hold-ups due to poor battery life and patent disputes—and a much
improved range 1 year later (see footnote 6), they will inevitably be expensive
for some time given short production runs which, in turn, reflect the variety of
different spectrum bands being used.

It has been estimated that to roll out a LTE network costs roughly €2 billion
per 50 million population covered (although the Aircom data for the USA cited
above appear to be more conservative), of which only a very small part—unlike
in the case of 3G—comes in the form of licence costs. In the specific case of LG
Uplus in South Korea, the cost of its nationwide roll-out in the 800 MHz band
was stated to be $1.1 billion (Sahota 2012). An investment on that scale cannot
currently be justified in many countries—especially those in less-developed
regions such as Africa—so roll-outs are likely to be limited in coverage for
several years to come although, as ever, it is very difficult (as was the case for

3G) for one operator to hold back when its main rivals have launched.[3] However, where HSPA+ already exists, loss of a 4G signal may not be all that obvious for the time being. It may be noted that 3G is now universal in developed countries and commonplace elsewhere, but it is easy to forget that a decade has passed since the first licences were issued and that little happened during the initial 3 years—Arthur D Little (2012), for one, argues that LTE is impacting much faster than UMTS.

It must also be remembered that a nationwide roll-out may take several years, and that in any event most operators do not intend their LTE networks to provide such coverage. What customers are going to find is that their devices flip from LTE to either HSPA+ or EDGE+ as they travel about. In this respect, it is also worth noting that operators are introducing LTE for a variety of motives. These include a desire to provide high-speed connectivity in rural areas, a desire to establish, or harden, a customer relationship via a reputation as a technical innovator and a desire to attract high-spending customers.[4] For their part, vendors view LTE as an opportunity to establish relationships with operators previously dependent on other vendors (although this is something of a zero-sum game).

However, one fundamental factor that has changed dramatically over the past few years has been the amount of data downloaded. The debate in the USA concerning the (alleged) shortage of spectrum has been noted above, and the ITU has stated that there is a need for between 1,280 MHz and 1,720 MHz of new spectrum to satisfy the demands of LTE-Advanced and subsequent developments. An ITU report in 2005 got its predictions for future data transfers hopelessly wrong, underestimating the maximum demand within 5 years by a factor of five. Recent data suggest a doubling of usage via mobile devices year-on-year (Cellular-news 2012d, j; Ericsson 2012) although some estimates are more conservative (Lennighan 2013). According to one study in 2012, LTE users were consuming 1.2 GB per month on average compared to 500 MB via slower technologies (Grant 2012). For its part, Verizon Wireless announced in October 2012 that whereas only 12 % of its contract customers owned LTE devices, they accounted for 35 % of its data traffic (Gabriel 2012h).

One interesting issue is the effect this will have on data plans, because the higher the download speed available, the greater the temptation to resort to data streaming which eats up data allowances at an alarming rate. Obviously, this tends to make 'all-you-can-eat' data plans uneconomic from an operator's perspective, and operators are beginning to experiment with new approaches such as shared plans taking in multiple users and services as initially introduced

---

[3] The prices paid for spectrum in South Korea elicited for the first time the fear that there would be a 'winner's curse'. This was predictably denied by the winners (TeleGeography 2011a).

[4] Although Detecon prefers to prioritise (1) securing capacity for new and faster services such as HDTV (2) use as fixed broadband wireless access and (3) expansion of capacity of existing networks (Cellular-news 2011a).

by Verizon and AT&T in the USA in June/July 2012 (Gryta 2012a). AT&T starting weaning its subscribers off unlimited data plans in September 2010 whereas Verizon removed that option for new subscribers in 2011. However, one regional operator that has launched LTE, but also committed to an unlimited data plan, is MetroPCS (Gryta 2012b),[5] and it may be noted that Sprint Nextel has yet to abandon its unlimited data plan—albeit in respect of WiMAX not LTE and costing $110 a month. For its part, T-Mobile USA returned to an unlimited data plan in September 2012 in respect of its HSPA+ service (Telecom.paper 2012e) at a cost of $90 a month with a subsidised handset or $70 a month without. It has been noted that, of the majors, T-Mobile was the only one not offering the iPhone (Luckerson 2012). However, in March 2013, T-Mobile USA announced that it would be launching the iPhone in April—having lost more than two million customers on contracts during 2012—and would be allowing customers to buy their own devices separate from the service contract (Gryta 2013).

Elsewhere in the world, the evidence as of October 2012 was that 4G data was being priced roughly 20 % higher on average than 3G. However, it was noted that because the introduction of LTE increases network capacity as well as transfer speeds, operators have some flexibility to either increase data limits or cut prices if they need to switch customers onto their 4G networks, and that this has already occurred on a number of occasions (Cellular-news 2012n).

If data downloaded continue to rise at the current pace, sooner or later there will undoubtedly be pressure on all operators to introduce faster downlinks. But the question is, what speeds will actually be delivered? As of the end of 2010, it was evident that so-called 'high-speed' networks rarely delivered more than 10 Mbps although that changed during 2011 as HSPA+ became more widespread.[6] As LTE is certain to be limited in geographical coverage for several years to come—at least in the great majority of countries—with networks falling back on HSPA+ where LTE is unavailable, it is worth asking whether customers will actually notice the loss of a LTE connection. In all probability, only those of a 'nerdish' disposition will bother checking what technology is being provided, and it must be emphasised that a 5–10 Mbps HSPA+ downlink will seem to be lightning fast compared to a year or so ago.

This suggests that sooner or later additional spectrum will have to be made available to accommodate the needs of LTE, let alone LTE-Advanced.[7] Over the

---

[5] A somewhat surprising finding in recent studies is that smartphone users often download massive files at great cost via their mobile operator rather than waiting until they are in range of a free Wi-Fi connection. The latter is particularly sensible for streaming video which is uses very large amounts of data (Grant 2012).

[6] According to Akamai Technologies, the highest average connection speed among 111 operators sampled at the end of 2010 was roughly 6 Mbps achieved by a Russian operator, whereas the highest peak speed of 23 Mbps was achieved by a Slovakian operator.

[7] There is an alternative, which is to use existing spectrum more efficiently. To get an idea of how this might be achieved see Cellular-news (2012c).

past decade, the European Union and its Commission has led the world in devising plans for harmonised spectrum, so it comes as no surprise that, in February 2012, the European Parliament approved a 5-year Radio Spectrum Policy Programme (RSPP). In principle—the Programme consists of guidelines that national regulators are requested to implement—the following key steps are to be taken: firstly, by the end of 2012, each Member State should have authorised the use of harmonised spectrum in the 2.6 GHz, 3.4–3.8 GHz and 900/1,800 MHz bands; secondly, by the same date but subject to derogation, each Member State should have authorised the use of the 800 MHz band; and thirdly, by 2015 at the latest, spectrum trading should have been permitted in the harmonised bands (Cellular-news 2012a).

So is LTE going to be a repeat of 3G? One thing is definitely different, which is that LTE licence fees have been fairly modest in the countries where, retrospectively, payments for 3G licences were considered to be excessive, although Italy is arguably an exception and it has not necessarily been the case in other countries as noted above. In addition, roll-out costs running, perhaps, to billions of dollars, but more often amounting to several 100 million, are much larger than those for HSPA+ upgrades. On the other hand, LTE is spectrally efficient and new sources of spectrum are definitely needed.

For the moment the jury must remain out, but it would be very surprising if LTE does not become the technology of choice for heavily congested networks by 2015, even if HSPA+ will run in parallel throughout the period. But this version of LTE will not be LTE-Advanced which will not come on stream on any significant scale prior to that date, even given final ratification of the qualifying IMT-Advanced technologies in early 2012, and it seems unlikely that the initial versions of LTE-Advanced will incorporate every possible item in the toolbox such as MIMO in the uplink and cognitive radio. So anyone expecting 'true 4G' in 2013 is in for a severe disappointment, although the vast majority of customers will not be aware of this given that, in a surprise decision made at the ITU World Radiocommunications Seminar 2010 (WRS-10) in December 2010, LTE, WiMAX and HSPA+ (at equivalent speed) could henceforth all be marketed as 4G (Cellular-news 2010; Wikipedia 2010).

As a result, at least for operators and vendors, Xmas 2010 brought an unexpected gift—one that operators in the USA immediately accepted. At the CES in early January 2011, AT&T switched the nomenclature of its existing 3G network to 4G, thereby completing a full set of nationwide operators claiming to have a 4G network (Cheng 2011). At the same time, vendors announced a variety of so-called 4G devices, for example the HTC Evo 4G Shift, the HTC Thunderbolt, the Motorola Droid Bionic, the Motorola Atrix 4G, the Samsung Galaxy S 4G and the 4G version of the BlackBerry Playbook. Samsung even went so far as to announce the Infuse 4G and Vibrant 4G, as did Motorola with the Sidekick 4G, designed to run exclusively on HSPA+ networks. The use of 4G to describe a HSPA+ network was first adopted in Europe by Netia P4 but, not to be outdone, AT&T went one stage further in May 2011 in claiming that—despite widespread agreement that any variant of 4G required a download speed of

21 Mbps—14.4 Mbps sufficed on its own network since it was accompanied by fast backhaul (Gabriel 2011a).[8]

As it happens, AT&T demonstrated in May 2011 that its LTE network would be capable of a 28.7 Mbps downlink, only for T-Mobile to announce a doubling of the downlink on its HSPA+ network to 42 Mbps. Naturally, both numbers are somewhat hypothetical, and neither operator specified the typical speed enjoyed by customers in the real world (Har-Even 2011a). An indication that the situation is getting out of hand is demonstrated by the claim by Rogers in June 2011 that its soon-to-be-launched LTE network would be capable of a 'theoretical' 150 Mbps downlink and 70 Mbps uplink which, in the light of its previous labelling of its 21 Mbps HSPA+ network as '4G', it would be promoting as 'Beyond 4G' (Har-Even 2011b).[9] An ironic twist was forthcoming in Liberia, where Cellcom advertised its new 4G service in July 2012 even though it was known to be a HSPA+ upgrade. The regulator launched an investigation as Cellcom had not been awarded a 4G licence, but it had clearly not breached its existing licence terms as it was licensed for HSPA and hence could be found guilty of nothing more than a touch of hyperbole (Cellular-news 2012l).

---

[8] In practice, as noted, 14.4 Mbps is technically HSPA rather than HSPA+, although it was rumoured that some devices contained a HSPA+ chipset. Thus, for example, the iPhone 4S, launched by AT&T in October 2011, would normally operate with HSPA but if it came within range of a HSPA+ signal it would be capable of faster speeds. Amusingly, it would simultaneously flag up '4G' on the screen.

[9] This trend has caused understandable unease, and in the USA a bill—'Next Generation Wireless Disclosure Act'—was introduced in June 2011 aimed at requiring operators to be more specific about their use of the term '4G' (Telecom.paper 2011). In an interesting case in South Africa, Cell C was fined for falsely advertising its 3G network using a 4G logo prior to the ITU decision of December 2010, but even though this practice had thereafter become permissible, the fine was upheld on appeal (Cellular-news 2011b). In contrast, in the UAE in November 2011, the regulator authorised Du to market its DC-HSPA+ network as 4G—alongside an admission as to the technology in use—even though Etisalat had already launched a LTE service (TeleGeography 2011b). For its part, C&W (branded as LIME) announced the roll-out of a so-called 4G network in Jamaica in March 2012 which appeared to be a 'refreshment' of its existing HSPA+ network (TeleGeography 2012a), and announced the launch of '4G Mobile' in June (Cellular-news 2012i). Of particular interest was the announcement by the Australian regulator in March 2012 that it intended to take Apple to court for breaches in consumer law. Specifically, it objected to Apple's use of the term 'iPad with Wi-Fi + 4G' in respect of a SIM card that could not, in practice, connect to any 4G network in Australia (Cellular-news 2012e) because it was set up only for spectrum used by AT&T and Verizon Wireless in the USA. Apple gave an undertaking to withdraw the advertising in March but failed to do so, leading to a court case which ended with Apple being fined A$2.25 million (Cellular-news 2012h). In practice, much the same situation took place in the UK (and elsewhere) during 2012H1, and Apple was forced to withdraw the term under pressure from the Advertising Standards Authority (Cellular-news 2012g). In a further interesting development, AT&T announced in July 2012 that it would no longer be referring to its HSPA+ network as 'America's largest 4G network' because all the major networks either had, or in its own case, shortly would have, launched LTE (TeleGeography 2012c). However, very few regulators have so far been willing to ban use of the term '4G' unless a service materially faster than that marketed as '3G' is being provided—but see Peru case study and TeleGeography (2012d).

Let us turn finally to a structural concern. We have observed elsewhere (Curwen and Whalley 2010) that there is a visible tendency for advanced markets to move towards an equilibrium where there are either three or four operators, and the likes of Orange and Vodafone appear keen on selling off their minority stakes without control. On the face of it, this process is not driven by technological progress, but that is not strictly true. In many cases, for example, the weaker operators in Europe delayed the roll-out of 3G because the cost did not appear to be justified in the light of the projected additional revenue from so doing, but this, in turn, provided a competitive advantage to those operators that moved early and established their brand. The same thing is likely to happen in the case of LTE. Just as with 3G, LTE is unlikely to promote new entry, and if the licences are acquired by the top three operators in each market, the trend towards concentration will be accelerated as customers switch to the fastest networks.

However, as is evident from, for example, the fierce opposition to the proposed take-over of T-Mobile USA by AT&T in the USA, regulators are unlikely to be enthused by the argument that the introduction of 4G justifies a reduction in the number of operators. It is increasingly likely, therefore, that a combination of high roll-out costs and regulatory issues will result in shared infrastructure for the provision of 4G services. In the USA, there may eventually be three networks, with all operators re-selling services, although it is not altogether easy as yet to determine who will own the networks (Gabriel 2011b). Meanwhile, in countries such as Belarus and Russia, no more than two networks are likely to be built.

References

4G-Portal. (2013). *Industry's smallest 4G LTE-Advanced modem from Broadcom.* Accessed February 12, 2013, from http://www.4g-portal.com

Arthur, D. L. (2012). *LTE spectrum and network strategies.* Accessed June 20, 2012, from http://www.adl.com/LTESpectrum

Cellular-news. (2010). *ITU declares that WiMAX and LTE can be classed as 4G networks.* Accessed December 23, 2010, from http://www.cellular-news.com

Cellular-news. (2011a). *Detecon defines success factors for 4G market launch.* Accessed March 1, 2011, from http://www.cellular-news.com

Cellular-news. (2011b). *South African network loses advertising appeal over its 4G branding.* Accessed August 31, 2011, from http://www.cellular-news.com

Cellular-news. (2011c). *NEC collaborating with Cisco to build commercial LTE networks.* Accessed September 1, 2011, from http://www.cellular-news.com

Cellular-news. (2012a). *European Parliament approves 5-year radio spectrum plan.* Accessed February 21, 2012, from http://www.cellular-news.com

Cellular-news. (2012b). *Ericsson leads LTE RAN standard contributions.* Accessed February 27, 2012, from http://www.cellular-news.com

Cellular-news. (2012c). *Pasta-shaped radio waves beamed across Venice.* Accessed March 2, 2012, from http://www.cellular-news.com

Cellular-news. (2012d). *Mobile internet usage is doubling year on year.* Accessed March 2, 2012, from http://www.cellular-news.com

Cellular-news. (2012e). *Australian regulator to seek orders against Apple for iPad 4G claims.* Accessed March 27, 2012, from http://www.cellular-news.com

Cellular-news. (2012f). *Option shows off quad-band LTE modem.* Accessed April 4, 2012, from http://www.cellular-news.com

Cellular-news. (2012g). *Apple drops 4G claim from new iPad following complaints.* Accessed May 14, 2012, from http://www.cellular-news.com

Cellular-news. (2012h). *Apple fined by Australian court for misleading 4G claims on its IPad.* Accessed June 25, 2012, from http://www.cellular-news.com

Cellular-news. (2012i). *Digicel launches HSPA+ upgrade in Jamaica.* Accessed June 29, 2012, from http://www.cellular-news.com

Cellular-news. (2012j). *Global mobile data traffic will reach 7 million terabytes per month by 2016.* Accessed June 29, 2012, from http://www.cellular-news.com

Cellular-news. (2012k). *Wireless infrastructure market at an 8-year low, at least two more quarters of pain.* Accessed July 20, 2012, from http://www.cellular-news.com

Cellular-news. (2012l). *Liberian regulator investigating Cellcom's 4G network claims.* Accessed August 7, 2012, from http://www.cellular-news.com

Cellular-news. (2012m). *Nokia Siemens networks and Huawei increased market share in Q2.* Accessed August 17, 2012, from http://www.cellular-news.com

Cellular-news. (2012n). *4G data is being priced 20% higher than 3G.* Accessed October 5, 2012, from http://www.cellular-news.com

Cellular-news. (2013a). *Qualcomm dominates LTE baseband chipset market, Broadcom and Intel fierce competitors.* Accessed May 5, 2013, from http://www.cellular-news.com

Cellular-news. (2013b). *Global LTE subscriber base to pass 100 million this week.* Accessed May 17, 2013, from http://www.cellular-news.com

Cheng, R. (2011). *'4G' takes centre stage at CES.* Accessed January 11, 2011, from http://www.totaltele.com

Curwen, P., & Whalley, J. (2010). *Mobile communications in a high-speed world: Industry structure, strategic behaviour and socio-economic impact.* Farnham: Gower Press.

Dignan, L. (2012). *Samsung vs. Apple: Are LTE patents the next battleground?* Accessed September 11, 2012, from http://www.cellular-news.com

Eogogicsinc. (2012a). *LTE operator strategies. Report R-1204D.* Accessed March 30, 2012, from http://www.eogogics.com

Eogogicsinc. (2012b). *LTE market by infrastructure, devices and operator services, 2012–2016. Report R-1209C.* Accessed September 20, 2012, from http://www.eogogics.com

Ericsson. (2012). *Traffic and market report: On the pulse of the networked society.* Accessed June 29, 2012, from http://www.ericsson.com

Gabriel, C. (2011a). *AT&T loosens android policies, and 4G definitions.* Accessed May 12, 2011, from http://www.rethink-wireless.com

Gabriel, C. (2011b). *Verizon stitches up rivals with shock cablecos deal.* Accessed December 5, 2011, from http://www.rethink-wireless.com

Gabriel, C. (2012a). *Renesas Mobile slashes cost of supporting LTE.* Accessed February 16, 2012, from http://www.rethink-wireless.com

Gabriel, C. (2012b). *Tele2 gets ready for LTE in Baltic states.* Accessed April 27, 2012, from http://www.rethink-wireless.com

Gabriel, C. (2012c). *Shortages loom in 4G basebands and advanced displays.* Accessed May 31, 2012, from http://www.rethink-wireless.com

Gabriel, C. (2012d). *Qualcomm and DoCoMo take different views on LTE roaming.* Accessed June 6, 2012, from http://www.rethink-wireless.com

Gabriel, C. (2012e). *VIA promises LTE patent pool at last.* Accessed June 15, 2012, from http://www.rethink-wireless.com

Gabriel, C. (2012f). *Antenova addresses LTE multiband challenge.* Accessed July 10, 2012, from http://www.rethink-wireless.com

Gabriel, C. (2012g). *Via licensing signs seven operators for LTE patent pool.* Accessed October 4, 2012, from http://www.rethink-wireless.com

Gabriel, C. (2012h). *Verizon's LTE network already handles 35% of its data.* Accessed October 10, 2012, from http://www.rethink-wireless.com

Gabriel, C. (2013a). *Marvell and Altair show off LTE device chips.* Accessed February 21, 2013, from http://www.rethink-wireless.com

Global mobile Suppliers Association. (2013a). *Report: Status of the LTE ecosystem.* Accessed February 2, 2012, from http://www.gsacom.com

Global mobile Suppliers Association. (2013b). *Report: Status of the LTE ecosystem.* Accessed May 8, 2012, from http://www.gsacom.com

Grant, K. (2012). *The smartphone data diet.* Accessed September 21, 2012, from http://www.arturogoga.com

Gryta, T. (2012a). *AT&T unveils shared mobile data plans.* Accessed July 19, 2012, from http://www.totaltele.com

Gryta, T. (2012b). *MetroPCS to launch 4G promotional plan Thursday.* Accessed August 22, 2012, from http://www.totaltele.com

Gryta, T. (2013). *T-Mobile USA unveils plans for iPhone, faster network.* Accessed 26, 2013, from http://www.totaltele.com

Har-Even, B. (2011a). *T-Mobile USA launches 42Mbps HSPA+ network upgrade.* Accessed May 24, 2011, from http://www.telecoms.com

Har-Even, B. (2011b). *Rogers to launch LTE in Ottawa this summer.* Accessed June 24, 2011, from http://www.telecoms.com

Lennighan, M. (2013). *Europe's mobile users make the most of 3G plans.* Accessed January 8, 2013, from http://www.totaltele.com

Lewis, M. (2012). *The death of embedded 4G.* Accessed March 24, 2012, from http://www.rethink-wireless.com

Luckerson, V. (2012). *Unlimited data plans: Are they coming back from the dead?* Accessed August 23, 2012, from http://www.lteworld.org

Lundborg, M., Reichl, W., & Ruhle, E.-O. (2013). Spectrum allocation and its relevance for competition. *Telecommunications Policy, 36,* 664–675.

Newlands, M. (2013). *UK 4G auction result embarrasses government.* Accessed February 20, 2013, from http://www.cellular-news.com

Ng, D. (2011). *Spectrum re-farming a cheaper alternative to buying LTE spectrum.* Accessed September 23, 2011, from http://www.cellular-news.com

Sahota, D. (2012). *Operators to invest as Korean LTE users want more.* Accessed August 29, 2012, from http://www.telecoms.com

Telecom.paper. (2011). *US bill proposal tackles 4G definition.* Accessed June 24, 2011, from http://www.telecompaper.com

Telecom.paper. (2012a). *SK integrates Wi-Fi, 3G, LTE networks to boost data speeds.* Accessed January 19, 2012, from http://www.telecompaper.com

Telecom.paper. (2012b). *ZTE to launch two LTE handsets at MWC.* Accessed February 21, 2012, from http://www.telecompaper.com

Telecom.paper. (2012c). *Marvell launches multimode TD LTE chipset.* Accessed March 2, 2012, from http://www.telecompaper.com

Telecom.paper. (2012d). *Hrvatski Telekom launches LTE in four cities.* Accessed April 27, 2012, from http://www.telecompaper.com

Telecom.paper. (2012e). *T-Mobile USA launches unlimited data plan.* Accessed August 22, 2012, from http://www.telecompaper.com

Telecom.paper. (2013). *Nvidia announces 2.3GHz Tegra 4i LTE smartphone processor.* Accessed February 20, 2013, from http://www.telecompaper.com

TeleGeography. (2011a). *SKT emerges as winner in race for 1800MHz spectrum.* Accessed August 31, 2011, from http://www.telegeography.com

TeleGeography. (2011b). *Du given green light to market HSPA+ service as 4G.* Accessed November 16, 2011, from http://www.telegeography.com

TeleGeography. (2012a). *LIME refreshes wireless sector with '4G' rollout.* Accessed March 13, 2012, from http://www.telegeography.com

TeleGeography. (2012b). *US remains at forefront of LTE service adoption.* Accessed March 20, 2012, from http://www.telegeography.com

TeleGeography. (2012c). *T-Mobile USA discontinues 'America's largest 4G network' tagline.* Accessed July 23, 2012, from http://www.telegeography.com

TeleGeography. (2012d). *Peru bans use of '4G' in marketing.* Accessed November 28, 2012, from http://www.telegeography.com

White, P. (2013). *Huawei pushes Scalable UMTS for refarming.* Accessed February 26, 2013, from http://www.rethink-wireless.com

Wikipedia. (2010). *4G.* Accessed December 1, 2010, from http://www.en.wikipedia.org/wiki/4G

Wood, N. (2012). *Vast majority of LTE devices support 700-MHz band.* Accessed April 10, 2012, from http://www.totaltele.com

Wood, N. (2013). *Android, iOS accounted for 87.6% of global smartphone market last year.* Accessed February 14, 2013, from http://www.totaltele.com

Making Use of Superfast Connectivity

9

9.1 Introduction

So far, this book has concentrated upon the direct relationship between 4G and consumers. However, the emergence of high-speed data transfers is transforming the world in which we live in ways that affect consumers only indirectly. The purpose of this chapter is to examine this phenomenon via a limited number of illustrative case studies.

The turning of attention towards the services that high-speed networks can deliver is not a recent phenomenon as it originated with the development of 3G. By consulting, among others, Curwen and Whalley (2010); Standage (2009), it becomes clear that mobile devices can be used to facilitate trade and provide banking, government and healthcare services. Some of these services have become well-established and generate significant revenues—near field communication-enabled mobile devices, for example, are increasingly common (Cellular-news 2013g, l) while the volume of mobile payments is forecast to reach $1 trillion by 2015 (Cellular-news 2013e).

However, given that the delivery and functionality of these and other services that already exist widely can only be improved through the use of 4G, they do not strictly present the kind of opportunity that would help to justify the considerable investment that mobile operators are making in their 4G networks. They represent, in effect, more of the same but faster. In contrast, it can be argued that substantial opportunities for mobile operators are to be found at the confluence of two key developments: Big Data and the Internet of Things.

Cukier and Mayer-Schoenberger (2013) argue that Big Data is transformational due to both the large amount of data that is now available and the uses to which it is being put. The amount of information that is available has grown rapidly in recent years. It was estimated that the global amount of information that had been created in 2005 was around 125 Exabytes (Cukier 2010) but in the space of just 4 years this figure increased sixfold to 750 Exabytes. While the volume of data is one characteristic of big data, Cukier and Mayer-Schoenberger (2013) argue that another is

P. Curwen and J. Whalley, *Fourth Generation Mobile Communication*,
Management for Professionals, DOI 10.1007/978-3-319-02210-9_9,
© Springer International Publishing Switzerland 2013

'datafication'—that is, turning parts of the world into data. They illustrate this through location, arguing that an individual's location is turned into data initially through longitude and latitude co-ordinates and, more recently, by the use of GPS.

While much of the discussion of Big Data in Cukier (2010) lies outside the scope of this discussion, a key issue that is relevant is that the analysis of Big Data identifies trends, insights and so forth that would not be possible with smaller datasets. Moreover, these insights are possible wherever large amounts of data can be compiled with the consequence that they are not limited to the 'traditional' areas such as the retail sector. The second key development noted above, the Internet of Things, will generate large amounts of data. Although various definitions of the Internet of Things have been proposed—see, for example, Fleisch (2010); International Telecommunication Union (2005)—although some have suggested that the concept is 'fuzzy' (Atzori et al. 2010; European Commission 2008)—the term is generally taken to refer to the blurring of the physical and electronic worlds (Coetzee and Eksteen 2011; Domingo 2012; Haller 2010; Siorpaes et al. 2006). This blurring occurs through the embedding of technology into devices, thereby enabling them to communicate with one another. In addition, the Internet of Things is characterised by the ability to identify devices and their interaction with one another (Miorandi et al. 2012; Kosmatos et al. 2011; Vermesan et al. 2009).

Radio-frequency identification is often associated with the Internet of Things (Darianian and Michael 2008; Gang et al. 2011; Haller et al. 2009; Zorzi et al. 2010). However, this has begun to change over time, reflecting both the conceptual broadness of the Internet of Things and a series of technological developments. Increasingly, 'smart objects' such as scanners, displays and embedded devices are discussed as part of the Internet of Things (Gang et al. 2011; Guinard et al. 2011; Kosmatos et al. 2011). Moreover, the areas where the Internet of Things can be applied have grown—for example, Atzori et al. (2010) suggests five broad application areas: transportation and logistics, healthcare, smart environments, personal and social and futuristic. Fleisch (2010) suggests that a series of value drivers—such as simplification and proximity—will result in value for businesses and consumers alike, leading to the widespread adoption of the Internet of Things.

The coming together of the Internet of Things and Big Data represents a substantial and potentially very lucrative opportunity. To realise this opportunity, co-ordination between a myriad of actors is required so that ideas are transformed into usable products and services (International Telecommunication Union 2005; Mainetti et al. 2011). Rather than focus on all of the areas where opportunities can be found, the remainder of this chapter will focus on two markets where mobile operators could play a role, namely in the machine-to-machine and smart cities markets. In both of these areas mobile operators can potentially leverage the speed and coverage associated with 4G to their advantage. In addition, both machine-to-machine (M2M hereafter) and smart cities are viewed as representing a substantial revenue opportunity for mobile operators. Nevertheless, mobile operators do face a series of challenges and obstacles that they will need to overcome if they are to compete successfully in both of these markets.

9.2   Machine-to-Machine

In their introduction to a special issue of IEEE Communications Magazine, Hu et al.
(2011: p. 24) provide an overview of M2M communication. While they do not
provide an explicit definition, they do assert that M2M communication is 'a form of
data communication between entities that do [sic] not necessarily involve any form
of human intervention'. Arab (2012) also mentions data, as does OECD (2012).
However, while data communication is central to M2M, other aspects can also be
found in the descriptions provided by Arab (2012); Hu et al. (2011); OECD (2012).

Hu et al. (2011) state that M2M communication is characterised by its low cost.
While they do not explain what is meant by 'low cost', either in absolute or relative
terms, Lawton (2004) suggests that standardisation and interoperability have
reduced the cost of implementation. Although the low cost of M2M devices is
expected to result in their adoption in very large numbers, some have argued that
this does not mean that all such devices will be characterised by low cost (Wu et al.
2011). Instead, a continuum is likely to emerge with devices ranging from low to
high cost on the one hand, and low to high performance on the other. It is also worth
noting that software developments have occurred (Niyato et al. 2011), improving
the intelligence and autonomy of devices. As a consequence, the range of functions
that can be undertaken by M2M devices has increased. In other words, M2M
devices both can be, and are being, put to an increasingly diverse range of uses.

Another characteristic of M2M communication is that the devices are connected
to one another, with Lawton (2004) among others suggesting that the value of an
individual connection is maximised when it is part of a larger network. A variety of
technologies can be used to connect M2M devices together. The choice of one
communication technology over another is not straightforward, involving a series
of trade-offs (OECD 2012). These trade-offs reflect the nature of the device itself,
where it is located, the sensitivity of the data collected and the urgency with which
the data collected needs to be transmitted. For example, a large number of devices
in close proximity and sharing the same resources would affect performance, while
lower power consumption by devices will influence their range. Moreover, if the
device moves, then it will need to work in a range of different environments. Given
these and other trade-offs, the communication technology facilitating M2M data
transfers could be either fixed or wireless. Possible fixed technologies include the
public switched telecommunications network (PSTN) as well as broadband
technologies such as DSL and fibre (OECD 2012), whereas wireless technologies
include Wi-Fi and successive generations of mobile technologies.

From our perspective, successive generations of mobile technologies are interesting for two reasons. Firstly, there are now more than five billion mobile
subscribers globally (OECD 2012). As the number of mobile devices has grown,
mobile network operators have invested considerable sums in their infrastructure.
When population distributions are coupled with coverage obligations imposed by
some regulatory authorities, the result is that the networks cover a substantial part,
if not all, of the population. In other words, mobile network are ubiquitous in many
countries, especially for 2G technologies (OECD 2012). However, as 2G-based

networks are decommissioned, 3G and 4G technologies will play an increasingly prominent role in M2M developments (Wu et al. 2011). Having said this, the smaller geographical footprint of 3G networks in many countries compared to 2G will mean that, geographically at least, once the latter is decommissioned the market will be smaller than it was (OECD 2012). Over time this is likely to change given that the attraction of using 4G technologies is that they are cheaper to implement as well as being spectrally more efficient.

The second reason why the use of successive generations of mobile technologies is of interest is that it provides a market opportunity for mobile operators to exploit. Not only could mobile operators provide wholesale access services to dedicated M2M service providers, but they could also enter the market themselves. While some mobile operators—such as Deutsche Telekom, Verizon and Vodafone—have entered the market themselves, it is clear that they will face many challenges both technically, economically and competitively.

Given the above, we can say that, for our purposes, M2M communication is characterised by data and low cost on the one hand, and by their ubiquity and use of wireless technologies on the other.

9.2.1 A Widely Applicable Technology

Where can M2M devices be used? Drawing on GSMA (2008) it is possible to identify a diverse array of areas where M2M devices could be utilised. A distinction is drawn between 'traditional' and 'non-traditional' M2M markets, with automotive applications and asset monitoring exemplifying the former and health, home applications and energy meters the latter. As can be seen from the table below, GSMA (2008) is not alone in indicating the wide array of areas where M2M technology could be employed (Table 9.1).

Zhang et al. (2011) explore how M2M networks could be implemented within the home environment. In doing so, they highlight the scope of application—from refrigerators to computing equipment, temperature control—as well as how the various technologies will need to be combined if the home M2M network is to be deployed. However, combining such technologies is not straightforward with the consequence that a series of challenges need to be overcome if the home M2M network is to be implemented. A broadly similar set of arguments can be found in Starsinic (2010), and although he argues that the combination of different technologies and devices together within the home M2M network presents certain challenges, it is suggested that a central role in integrating the various technologies will be played by the M2M gateway for the home.

Smart homes are grouped together into what Li et al. (2011) refer to as a 'smart community', namely, homes close to one another networked through wireless technologies such as Wi-Fi and 3G. They suggest that the networked platform that results can support the development of various services, with perhaps one of the more interesting being that of 'neighbourhood watch'. Individual homes within the community are fitted with surveillance cameras, which identify suspicious events

Table 9.1 Possible uses for M2M devices

Source	Areas of possible M2M use
Bartoli et al. (2010)	Smart grids
Cellular-news (2012i)	Vending machines
Evans and Annunziata (2013)	Aviation, power, healthcare, rail, oil and gas
Fadlullah et al. (2011)	Healthcare, vehicles, smart homes, smart grid
Foschini et al. (2011)	Road traffic management
GSMA (2008)	Consumer electronics, clean technology, health, transport and utilities
Kim and Song (2008)	Manufacturing
Nique and Arab (2013)	Energy, water, healthcare
Siegele (2010)	Smart meters, smart grids, water
OECD (2012)	Smart meters, vehicles, health, factory automation, consumer electronics, logistics, monitoring
Wu et al. (2011)	Remote maintenance and control, healthcare, vehicles, smart grid, tracking and tracing

and determine whether they pose a threat. Details of these threats can be shared with the network of neighbouring houses and, if necessary, a central hub. Naturally, implementing such a system is complicated. Threatening events need to be distinguished from those that are not, and computational resources need to grow as more events are tracked and assessed.

While the challenges identified and discussed by Zhang et al. (2011) emanate from a relatively narrow application of M2M technologies, the challenges identified by Lu et al. (2011) are considerably broader. They argue that the widespread adoption of M2M requires that the challenges associated with energy efficiency, reliability and security are overcome. The energy efficiency challenge is derived from the large number of M2M devices in use and their need to communicate, while the reliability challenge reflects the need for all of the various technology components to work. The security challenge relates both to the need to protect the operations of M2M devices from external interference, but also the need to ensure that the data collected are securely stored. Although suggestions are made with respect to all three areas, they conclude that more research is required.

M2M technology can also be applied to various forms of monitoring. Cristaldi et al. (2005) investigate the application of M2M technology to environmental monitoring, suggesting the hardware to be used in the process, while Nique and Arab (2013) show the benefits of monitoring in terms of improved service provision and reliability. Interestingly Nique and Arab (2013) identify such benefits within emerging markets—they state that as a result of monitoring water consumption in Kenya, new business models have emerged that address the maintenance needs of the widespread use of water pumps. The benefits of monitoring to the consumer in the sustainable energy example that they recount are less clear-cut, although the monitoring to determine failure rates could arguably lead to the improved reliability of supply. Monitoring is also integral to the application of M2M technology to road traffic management that is explored by Foschini et al. (2011). For example, the technology could be used to limit access by drivers to different parts of a city depending on the time of day or amount of traffic on a particular road. That said, the application of M2M technology is not without its problems—not only are there technical issues to overcome, but there is a need to ensure that M2M technology is integrated with the other information system-based management approaches. In other words, if the benefits of M2M technology are to be maximised, the data and analysis need to be shared with others.

One area of M2M application that has attracted considerable attention is that of smart grids, with Niyato et al. (2011) going as far as to argue that smart grids are one of the main drivers of M2M growth. Smart grids involve the application of M2M technology across the energy sector (Bartoli et al. 2010; Tan et al. 2011), with the collected data enabling, among other things, both energy efficiency and the reliability of the system to be improved (Li et al. 2011). While many focus on the benefits to the operators of energy networks, Bartoli et al. (2010) also suggests that consumers will benefit as well. On the one hand, consumers could receive their bills almost instantaneously, while on the other hand they could shift their energy consumption patterns to take advantage of lower tariffs. Liu et al. (2011) also suggest, perhaps speculatively, that smart grids can play a role in the use of electric vehicles through monitoring electricity consumption as well as the performance of the vehicle itself.

For these benefits to be realised, M2M technology needs to be applied across the whole energy sector of generation, transmission and consumption. While the increasing emergence of smart grid projects around the globe suggests that there is considerable interest in them, it is clear from Fadlullah et al. (2011); Li et al. (2011); Niyato et al. (2011); Tan et al. (2011) that the development of smart grids is technically challenging. Not only does a broad array of technologies have to be integrated together, so that, for example, data flow across the network, but this also has to occur cost effectively and reliably. Moreover, the data also needs to be secure, with Bartoli et al. (2010) suggesting that ensuring data security through aggregation techniques is possible but difficult to achieve.

Healthcare is another area where M2M technology could be applied. A broad array of healthcare applications is implied in GSMA (2008), from overweight individuals to monitoring patients with chronic diseases. Wu et al. (2011) suggest that the technology could be used to monitor patients remotely, with the patient's doctor or hospital being automatically updated as changes occur. In a similar fashion, Doukas et al. (2011) suggest that a range of technologies could be developed that would enable the elderly and disabled to live at home for longer than would otherwise be the case. Alternatively, the technology could be used within ambulances en route to hospitals to update them as to the medical condition of the patients being transported (Wu et al. 2011).

M2M technology could also be used to help provide healthcare to those located in rural areas. Nique and Arab (2013) mention a GPRS-based service that allows for test results from laboratories to be swiftly delivered to remote clinics. While M2M technology could be used to improve how healthcare is delivered, it could also be used in other ways (Li et al. 2011), for example to facilitate care in the community programmes with the sensors monitoring the patient and directing an appropriate response if a problem is detected. Significantly, if the M2M system is aware of the specific medical circumstances of a patient, the response could be tailored accordingly. This raises the possibility not only of more effective medical treatment, but also of an improvement in the efficient use of resources.

9.2.2 How Large Is the Potential Market?

Although many commentators argue that the M2M market represents a substantial opportunity for mobile operators, estimates of the size of this market are very varied. As M2M is often discussed in the context of devices being connected to the Internet, it is not unreasonable to imagine a market of several 100 million or even more connections. Haifei (2010) asserts that there will be 'tens of billions' of communication-enabled devices as more devices are manufactured, whereas Wu et al. (2011) suggest that, due to the ubiquitous nature of the devices, the number of connections would be somewhere between 'billions and trillions'. Both Arab (2012); OECD (2012) state that 50 billion connections are predicted by 2020, with the latter suggesting that the number will grow tenfold in the 8 years from 2012 to 2020.

A more modest number of connected devices by 2020 can be found in Cellular-news (2012a), which uses a report from Adaptive Media to suggest that there will be five billion machine-to-machine devices by 2020. A 2011 report from Analysys Mason, cited in Cellular-news (2011a), suggests that there will be 2.1 billion machine-to-machine devices connected by 2020. There are also a series of estimates considerably lower than these. A report from Juniper Research, cited in Cellular-news (2012p), predicts just 400 million connections by 2017, while ABI Research suggests 450 million cellular connections by 2018 (Cellular-news 2013m). IMS Research, cited in Cellular-news (2012s), forecasts 326 million cellular M2M connections by 2016. Interestingly, in the previous year, IMS Research forecast 100 million cellular M2M connections by 2015 (EE Times 2011).

The same report by Analysys Mason also suggests that the revenue generated by each connected device will vary between $0.25 and $40 per month. When the number of devices is combined with such revenue possibilities, it is understandable that some have suggested that the M2M market will generate substantial revenues Cellular-news (2011d). Juniper Research, cited in Cellular-news (2011b), estimates that the market will be worth $35 billion by 2016 whereas the considerably larger figure of $948 billion by 2020 has been suggested by Machina Research (Taylor 2013). Not to be outdone, the GSMA has forecast that the growth in wireless connected devices, which include M2M connections, will result in a $1.2 trillion market opportunity for mobile operators by 2020 (Cellular-news 2011e). It has also been suggested that M2M revenues for mobile operators will surpass those from voice by 2018 (Cellular-news 2013f), with revenues coming from both developed and developing countries and a variety of different sectors.

Finally, it is worth briefly noting the size of the market opportunity identified by Evans and Annunziata (2013). In their discussion of what they refer to as the 'Industrial Internet', and which shares similarities with both the Internet of Things and M2M communication, Evans and Annunziata (2013) identify where it could be applied and its potential socio-economic impact. The Industrial Internet could be applied to the industrial and non-industrial sectors of advanced and developing economies. Given a global gross domestic product of approximately $70 trillion in 2011, opportunities exist in those parts of the global economy that collectively amount to $32.3 trillion (Evans and Annunziata 2013). There is, in other words, significant scope for the application of the Industrial Internet. By arguing that the Industrial Internet could improve productivity in various sectors, Evans and Annunziata (2013) identify some of the savings that are possible—for example, if its application to the aviation sector saved 1 % of fuel this would amount to $30 billion over 15 years, while a 1 % reduction in healthcare inefficiency would equal $63 billion over the same period. Although Evans and Annunziata (2013) do not explicitly identify the size of the market that would result from its application, it is reasonable to assume that given how extensively it could be applied and the magnitude of the possible savings that the market is likely to be worth tens of billions of dollars.

These figures provide some indication as to the potential size of the M2M market. While this market is clearly attractive to mobile operators (Cellular-news 2012c), the presence of other companies providing M2M products and services inevitably means that their connections and revenues will be smaller than some of the eye-catching figures noted above. While the number of cellular M2M connections may indeed be large, with one consultancy forecasting 2.1 billion connections by 2020 (Morrish 2012) and another suggesting 2.5 billion connections by the end of 2020 (Cellular-news 2012m), the actual number of M2M connections being reported by mobile operators suggest that growth may not be as fast as many have anticipated.

But how many M2M connections do mobile operators actually have? From Cellular-news (2012f), it would appear that mobile operators have a relatively modest number of M2M connections, especially when compared to some of the larger forecasts noted above. At the end of the first quarter of 2012, AT&T had 13.3 million M2M connections while Sprint had 3 million connections and Softbank 1.9 million. This would suggest that mobile operators have not been that successful in expanding into the M2M market. However, Cellular-news (2012f) suggests that this picture, of relatively modest success by mobile operators in the M2M market, is slightly misleading for three reasons.

Firstly, some mobile operators such as Vodafone and Deutsche Telekom report the number of their M2M connections for only some of their subsidiaries. Thus, Vodafone has more than the 7.8 million connections reported in Cellular-News (2012d), and Deutsche Telekom more than the 3 million reported by T-Mobile USA. Secondly, some operators do not report separately the number of their M2M connections (Cellular-news 2012f). China Mobile, for example, is one such company and it was estimated to have 14 million M2M connections at the end of 2011

(Berg Insight 2012). Thirdly, on some occasions the forecast of M2M connections includes all technologies while on other occasions one specific technology, such as cellular, is specified. In other words, the forecasts are often not comparable.

The relatively large number of M2M connections achieved by China Mobile also highlights another feature of this developing market, namely, that emerging markets are central to its growth. Cellular-news (2011g, 2012e) cites research by consultancy companies that demonstrate the potential of the Asia-Pacific region in general and of China in particular. Pyramid Research, cited by Cellular-news (2011g), forecasts that Asia-Pacific will account for almost 40 % of the global M2M market by 2016. Having said this, it has been suggested that the lack of a specific M2M regulatory framework is hindering the development of the market in Asia-Pacific countries (Cellular-news 2012b; Detecon Consulting 2012). The obstacles identified by Detecon Consulting (2012) are, in fact, quite broad in their scope—switching costs, roaming, numbering—and echo calls by others for action in areas such as numbering and the wider legal framework (European Communications Committee 2010; OECD 2012; Weber 2009).

9.2.3 Value Chain

In many respects, the M2M value chain is relatively simple. If we draw on the likes of Lawton (2004), the value chain consists of a machine that is connected to a physical entity which, in turn, transmits data through a communications technology to another machine. These data are then used to provide a particular service. A more detailed value chain can be found in OECD (2012), where a clearer distinction is made between the various stages. A stylised version of the value chain that they suggest is as follows: Consumer to M2M device to network to M2M provider to application.

Haifei (2010); Lien et al. (2011); Wu et al. (2011) show that the M2M value chain is highly complex. Wu et al. (2011) use the term 'ecosystem' instead of value chain to reflect the dynamic interaction of a large number of stakeholders across many different industries. However, they do not provide a detailed description of how the various parts of the ecosystem interact with one another, focusing instead on identifying the challenges that will emerge once a large number of M2M devices are in use. In contrast, although Haifei (2010) does not explicitly identify a M2M value chain, he does argue that it is complex because of the number of actors involved in the provision of all the various components that go together to provide the service. Interestingly, Haifei (2010) notes that some mobile operators have developed their own M2M middleware software. Given the costs associated with such developments and the pace of change within the market, it is suggested that mobile operators will initially form partnerships and then acquire other companies in the sector to bolster their competitiveness.

Reports in Cellular News demonstrate that mobile operators are forming partnerships and acquiring other companies, with some of the recent examples as follows:

- France Télécom and Deutsche Telekom entered into a wide-ranging collaborative agreement in February 2011 to develop their M2M businesses. This agreement initially covered four European countries, but was extended, at least in part, in recognition that connectivity plays a key role in the M2M market. The geographical scope of the co-operative agreement was further expanded when TeliaSonera joined in July 2011 (Cellular-news 2011f).
- Telefónica and China Unicom agreed to develop M2M services in October 2011 (Cellular-news 2011f). Not only did the agreement entail the delivery of these services through a broad array of technologies, but also stipulated that the two companies would collaborate in the development of M2M platforms.
- In June 2012, Verizon paid $612 million to acquire Hughes Telematics. This acquisition expanded Verizon's presence in the automotive and fleet machine-to-machine markets (Cellular-news 2012h). This purchase is also mentioned by Taylor (2013), who also notes that AT&T has acquired Xanboo and Telefónica has purchased Masternaut.
- Seven mobile operators—KPN, NTT DoCoMo, Rogers Wireless, SingTel, Telefónica, Telstra and VimpelCom—formed an alliance to develop their M2M businesses in July 2012 (Cellular-news 2012k). Through this alliance the operators will collaborate to develop an infrastructure to enable M2M devices to be connected seamlessly and globally.
- The Open Mobile Alliance joined in September 2012 the oneM2M Partnership, another alliance that was launched a few months earlier to facilitate the deployment of M2M services (Cellular-news 2012n).
- In May 2013, Telenor agreed to provide M2M services to customers of KDDI, a Japanese telco (Cellular-news 2013i). A key feature of the agreement was that KDDI gained access to Telenor's roaming agreements in more than 190 countries.

While these developments illustrate that mobile operators are beginning to develop their M2M businesses, they shed light on only some of the developments that have been occurring. At the moment, geography clearly matters to mobile operators. By expanding their geographical coverage, through collaborative arrangements and joint ventures, mobile operators are recognising that, in the future, M2M devices will be mobile and that they need to be able to collect data from these devices regardless of their own footprint. That said, Cellular-news (2012j) suggests that geographical scope alone will not determine the profitability of the M2M business of mobile operators.

Mobile operators are also recognising the benefits of M2M platforms. There are undoubted economic and efficiency gains associated with the use of a single platform, especially when the platform covers a large geographical footprint. There are also likely to be some competitive advantages accruing to the mobile operator leading the development of the platform that is subsequently adopted by other companies, perhaps in terms of licensing payments but also in terms of being better placed than others when it comes to understanding how the technology works. Interestingly, Cellular-news (2013h), drawing on research by Heavy Reading Insider, suggests that through the in-house development of M2M platforms by mobile operators they are finding ways to reduce costs and thus position themselves in the market. In other words, collaboration is only one of the strategies that is emerging as mobile operators seek to position themselves in the M2M market.

Equipment vendors such as Ericsson appear to be actively developing a presence in the M2M market. Not only has Ericsson sold its own products to operators such as TeliaSonera (Cellular-news 2012r) or provided consultancy services to others like MegaFon (Cellular-news 2012q), but it has also sought to develop new services through entering into collaborative arrangements with a range of different companies. For example, Ericsson has signed collaborative deals with Pacific Controls, a software developer, to develop new M2M applications for a variety of markets (Cellular-news 2011c) as well as agreed to sell cloud-based M2M services jointly with SAP (Cellular-news 2013d). Ericsson has also started collaborating with another software company to simplify M2M device management (Cellular-news 2013b) which, if successful, should contribute to the wider adoption of M2M technologies.

Other equipment vendors apart from Ericsson are also expanding their presence in the M2M market. NEC agreed with Wyless, the developer of a M2M platform, to jointly explore the development of new services (Cellular-news 2012c), whereas Dell signed a pan-European deal with Telefónica (Cellular-news 2013j). Wyless also signed a deal with Telefónica, although the emphasis in this case was on expanding its geographical footprint through gaining access for its customers to Telefónica's networks (Cellular-news 2013k). Qualcomm has sought to leverage its position as a provider of chips to enter the M2M market (Cellular-news 2012e, o), providing new software as well as a clear roadmap of how its chips will develop in the future. Rather unusually, Sony formed a joint venture—Convida Wireless—with InterDigital, a patent holding company (Cellular-news 2013a). The joint venture will be funded by its parent companies to engage in M2M platform research and development (Sony Corporation of America 2013).

The increasing presence of Ericsson and other vendors clearly highlights the attractiveness of the M2M market, but it alludes to its heightened complexity as well. This is further compounded when the presence of other equipment vendors and M2M specialists is taken into account. One such specialist is Jasper Wireless, a company that provides software to mobile operators to enable them to deliver their own M2M services (Cellular-news 2012g, t, 2013c). Jasper Wireless has clearly been highly successful in selling its software to mobile operators—its website shows that operators around the globe are using its software and, in late 2012, it was ranked the leading provider of 'Connected Device Platform' software (Cellular-news 2012t). Whether by design or by accident, one advantage of being the leading provider of such software is that mobile operators in different countries will be able to collaborate to roll out services globally to their customers as a consequence of using the same underlying platform.

9.3 Smart Cities

Although the application of M2M technology to smart areas has been mentioned above, it is worth elaborating on 'smart cities' given the rapid urbanisation that is being experienced in many countries and the opportunities that this presents for mobile operators (GSMA 2011, 2013b). While the notion of 'smart cities' is arguably not a new one, having been around in various guises since the early 1980s (Batty 2012), the idea has come to prominence in recent years as governments struggle to cope with population growth, shifting patterns of economic activity, pollution and declining resources (GSMA 2011; Lee et al. 2013; Siegele 2010; Zhao et al. 2013).

Lee et al. (2013) argue that the 'smart city' concept is different from other concepts such as an 'intelligent city' through its focus on human capital and education, whereas Siegele (2010) suggests that common to the smart city initiatives that are underway is their desire to improve the management of the urban environment. Rather than offer a definition of a smart city, GSMA (2011: p. 5) instead suggests four common principles that should characterise a smart city. These are:

- That access to 'enhanced information flows' is provided to individuals and companies.
- Different data sets are combined to identify areas where productivity gains are possible.
- The achievement of economies of scale and scope across different parts of the city's infrastructure through the use of a common platform.
- Innovative technology and innovation are used by the city to improve its sustainability, local environment and the quality of life of its population.

Given these common principles, it is no surprise that a diverse array of activities fall within the scope of a smart city. Broadly speaking, however, these activities arguably fall into one of four categories: transport, environment/energy, municipal activities and economic development and open data (GSMA 2013b). Smart city developments within each of these four areas inevitably bring together a series of actors, which raises the issue as to the role that mobile operators could play. One possible role relates to the provision of infrastructure and thus connectivity (GSMA 2011, 2013b). Through a combination of 2G, 3G and increasingly 4G networks, mobile operators provide extensive coverage. However, mobile operators are not the only possible providers of connectivity—Rezende et al. (2013), while outlining a digital city project in Brazil, demonstrate the role played by fixed (fibre) infrastructure, whereas Middleton and Bryne (2011) explore whether Wi-Fi hotspots could be shared to create a wireless broadband infrastructure. Moreover, while the provision of connectivity alone would undoubtedly generate revenues, it is unlikely to do so on the scale that many had predicted (see above) or that offsets the decline in other revenue sources such as voice and international roaming.

Thus, it is likely that mobile operators will actively participate in the other opportunities identified by GSMA (2011, 2013b). Mobile operators could aggregate data from across the city and engage in its analysis, thereby providing some form of value-added to their clients. It is debatable whether mobile operators acting alone posses the necessary skills and competences to do this, with the consequence that they may form joint ventures with specialist companies. Perhaps a more likely scenario will see mobile operators focus their smart city efforts on the two other areas of opportunity identified by GSMA (2013b), namely, service delivery and customer interface. By their very nature, mobile operators provide a range of communication services to their customers and manage their infrastructure accordingly. This would suggest that they possess some, if not most, of the relevant skills and competences necessary to collect and distribute data on a real time basis. These skills and competences could be applied to pivotal smart city activities such as transport or energy. GSMA (2013b) also suggests that mobile operators could support smart city initiatives through, among other things, their call centres or raising awareness of such initiatives with their subscribers. The provision of call centre services may generate some revenues for mobile operators, though similar to connectivity, this is unlikely to offset declines elsewhere.

One area where mobile operators could play a role in is that of disaster management. While disaster management is a difficult and complex task, as can be seen from, for example, Aloudat et al. (2013); Dorasamy et al. (2013); Pauchant and Mitroff (1990), mobile operators can play a key by virtue of their widespread coverage. GSMA (2013a), which focuses on Japan and the 2011 tsunami, highlights the role played by mobile operators. Not only were the mobile operators involved in the planning of the system, but once the earthquake was detected they also began to implement the disaster management plan that had been agreed. The major Japanese mobile operators sent a warning text to the subscribers, while subsequent to the earthquake they began to restore services and ensure the resilience of those that survived. They also began to provide services that would help deal with the consequences of the earthquake and assist with the recovery—for example, mobile base stations were deployed.

A broader examination of the role of mobile technologies in disaster management can be found in Aloudat et al. (2013), who explore the role and acceptance of location-based government services. Using survey data from New South Wales in Australia, they found that the perceived usefulness of the services to an individual was a key factor in determining use of the service, and that individuals can evaluate the usefulness of the service if they can observe it. They also found that privacy concerns were expressed by respondents, with the collection of personal information resulting in trust in the service and its perceived usefulness being reduced.

9.4 Overcoming Challenges

M2M and smart cities represent areas of opportunity for mobile operators. Although estimates of the size of the two markets vary, sometimes quite considerably, they are clearly large and offer the chance for mobile operators to generate new revenue streams. These new revenue streams will help fund the considerable investment in 4G infrastructure that is needed and offset declines in other markets such as voice telephony. This assumes, however, that mobile operators possess all the necessary expertise and resources to enter and then successfully compete in both markets. Inevitably this raises the question as to whether or not this is the case.

Telecom.com (2013) asserts that mobile operators lack the necessary skills and resources to prosper in the Internet of Things. Several reasons are suggested for this, the first of which is that the telecommunications and M2M markets are not the same. Over 30 or so years, mobile operators have created a global market that enables users to call one another. In contrast, the M2M market is not a single market but rather one composed of many niche markets, determined by location and sector, each with different characteristics. Moreover, the financial characteristics of the two markets also differ—the average revenue per user (ARPU) in mobile telecommunications is claimed to be $200 per year while in the Internet of Things it is between $4 and $6. Although this implies that expanding into the M2M market at the expense of mobile APRU would be a 'pyrrhic victory', it is not clear why this would be the case but presumably it is because the margins for M2M are lower than those for mobile telecommunications.

Secondly, in M2M communications the device connects to a database within an organisation. These organisations are making long-term investments in M2M technologies and devices and need to be certain that the mobile operator and the services that it currently provides will be available throughout that period. It is argued that mobile operators are expanding into M2M as a way of utilising spare capacity, with the result that it is not certain that they will still be active in the market in the future when they face capacity constraints.

The third reason suggested is that business models will change in the coming years. One aspect of this is the method of allocating spectrum although this is not spelt out. More substantively, the related point is made that other companies such as Microsoft, IBM and Google, with an existing interest in Big Data and business-to-business services, may enter the market. Finally, it is suggested that technological bifurcation may occur. While this point is unclear, it seems to imply that 4G and M2M technologies will co-exist with one another.

The issues raised interact with one another to create a situation whereby mobile operators are unable to compete in the new markets that are emerging. However, the brevity of some of the points articulated necessitates that they are treated with a degree of caution. For example, there is limited elaboration of the relative financial attractiveness of the mobile and M2M markets or the technological bifurcation that could occur. However, a recent report would appear to support the argument that the M2M market is not attractive financially to mobile operators. According to Cellular-news (2013m), M2M ARPU is declining. This can be explained in part by the competitive nature of the market, and partly by the limited use of 3G and 4G technologies to deliver M2M connections. Although the margins for 3G and 4G-based connections are better than those for 2G, the majority of connections in 2012 used the latter technology.

Although Telecom.com (2013) mentions just three potential entrants into the M2M market, it is easy to imagine a diverse array of other companies entering the market –for example, SAP as well as consultancies, retailers and logistical companies. Indeed, any company collecting and then analysing data through M2M could conceivably compete against the mobile operators as they would have the necessary expertise gleaned from their own activities. That said, the degree to which they could compete in other sectors may be limited.

So far, a rather bleak picture of the ability of mobile operators to compete in the M2M and smart cities markets has been painted. There are, however, several reasons to suggest that mobile operators could actively compete in this market. In general, mobile operators are well-resourced companies. The lucrative nature of the mobile market means that mobile operators could adopt a variety of strategies to expand their presence in the markets. They could, for example, compete on price. This strategy could increase the overall size of the market as well as facilitate the achievement of scale economies and technological improvements that maintain or even improve margins. Moreover, if mobile operators did decide to compete on price they could conceivably undermine the viability of some of their rivals, forcing them out of the market and thus opening up the possibility in the future of price increases as competitive pressures decline.

Alternatively, mobile operators could partner with, or purchase outright, those companies with complementary but necessary resources that would enable them to compete in the markets. Cellular-news (2013m) suggests that the value added to consumers in M2M markets is provided through the provision of applications, systems integration and data analysis. If mobile operators were to adopt a strategy of acquiring access to resources in order to compete, the purchases should arguably occur in these areas. In essence, such a strategy acknowledges that the existing competitiveness of the mobile operators is derived from their networks. One aspect of this competitiveness relates to the deployment and operation of mobile networks. As this involves mobile operators working with a multitude of other companies such as equipment vendors, it demonstrates their collaborative abilities.

Secondly, mobile operators run networks in one or more countries. The geographical footprint that results enables mobile operators to deliver a range of services, M2M included, to their customers. Through continued and substantial investment in these networks, mobile operators are able to support a wide array of services and improve the quality of service that is being delivered. With the move towards 4G, more bandwidth is available—this bandwidth can be used to deliver 'traditional' services like voice telephony as well as more data intensive ones like M2M. Although there are some (technical) challenges involved in the delivery of multiple services over the same network without compromising quality, the clear advantage of such a strategy is that it maximises the revenues accruing to the mobile operator.

The M2M market has also witnessed mobile operators collaborating with one another. Alliances and joint ventures have been formed that extend the geographical footprint of individual operators. Without such collaborative arrangements, mobile operators will not be able to provide services in M2M markets such as logistics, aerospace or automotive monitoring that are inherently international in scope and wider than the footprints of most mobile telecommunication companies.

The collaboration has also sought to address the technical challenges associated with M2M, in terms of agreeing standards and developing appropriate platforms.

References

Aloudat, A., Michael, K., Chen, X., et al. (2013). Social acceptance of location-based mobile government services for emergency management. *Telematics and Informatics*, forthcoming.

Arab, F. (2012). *The other M2M opportunity: Enhanced utility access in emerging markets*. Accessed October 4, 2013, from http://www.gsma.org

Atzori, L., Iera, A., & Morabito, G. (2010). The internet of things: A survey. *Computer Networks, 54*, 2787–2805.

Bartoli, A., Hernandez-Serrano, J., Soriano, M., Dohler, A., Kountouris & Barthel, D. (2010). *Secure lossless aggregation for smart grid M2M networks*. Accessed June 11, 2013, from http://www.ieeexplore.ieee.org

Batty, M. (2012). Editorial – smart cities, big data. *Environment & Planning A, 39*, 191–193.

Berg Insight. (2012). *M2M research series – the global wireless M2M market*. Accessed July 27, 2013, from http://www.berginsight.com

Cellular-news. (2011a). *Machine-to-machine connections to reach 2.1 billion devices by 2020*. Accessed February 16, 2011, from http://www.cellular-news.com

Cellular-news. (2011b). *M2M to generate $35bn in service revenues by 2016*. Accessed May 17, 2011, from http://www.cellular-news.com

Cellular-news. (2011c). *Pacific controls signs M2M collaboration with Ericsson*. Accessed August 25, 2011, from http://www.cellular-news.com

Cellular-news. (2011d). *Cellular M2M subscriptions to increase almost fourfold between 2010 and 2016*. Accessed October 1, 2011, from http://www.cellular-news.com

Cellular-news. (2011e). *The $1.2 trillion revenue opportunity for mobile operators by 2020*. Accessed October 11, 2011, from http://www.cellular-news.com

Cellular-news. (2011f). *Telefónica signs M2M cooperation agreement with China Unicom*. Accessed October 20, 2011, from http://www.cellular-news.com

Cellular-news. (2011g). *Asia-Pacific to become largest M2M market in 2013*. Accessed November 3, 2011, from http://www.cellular-news.com

Cellular-news. (2012a). *Machine to machine technology adoption set to explode by 2015, but are we ready?* Accessed February 27, 2012, from http://www.cellular-news.com

Cellular-news. (2012b). *M2M growth in southeast Asia in need of regulatory incentives*. Accessed March 19, 2012, from http://www.cellular-news.com

Cellular-news. (2012c). *NEC and Wyless agree to cross-promote M2M services*. Accessed April 18, 2012, from http://www.cellular-news.com

Cellular-news. (2012d). *Global M2M subscriber base now exceeds 100 million*. Accessed April 19, 2012, from http://www.cellular-news.com

Cellular-news. (2012e). *Qualcomm announces roadmap and resources supporting M2M services*. Accessed May 8, 2012, from http://www.cellular-news.com

Cellular-news. (2012f). *AT&T tops estimates for largest M2M subscriber base*. Accessed May 25, 2012, from http://www.cellular-news.com

Cellular-news. (2012g). *O_2 signs M2M management deal with Jasper Wireless*. Accessed May 31, 2012, from http://www.cellular-news.com

Cellular-news. (2012h). *Verizon boosts M2M service capabilities with Hughes Telematics acquisition*. Accessed June 2, 2012, from http://www.cellular-news.com

Cellular-news. (2012i). *Cellular M2M connections in the retail industry surpassed 10 million in 2011*. Accessed June 4, 2012, from http://www.cellular-news.com

Cellular-news. (2012j). *Connectivity remains the focus of M2M telecom service providers in Europe*. Accessed July 4, 2012, from http://www.cellular-news.com

Cellular-news. (2012k). *Seven mobile networks form new M2M alliance.* Accessed July 10, 2012, from http://www.cellular-news.com

Cellular-news. (2012l). *M2M connections lags behind industry expectations.* Accessed September 21, 2012, from http://www.cellular-news.com

Cellular-news. (2012m). *M2M cellular connections expected to hit 2.5 billion by 2020.* Accessed September 21, 2012, from http://www.cellular-news.com

Cellular-news. (2012n). *Open mobile alliances joins M2M partnership.* Accessed September 26 ,2012, from http://www.cellular-news.com

Cellular-news. (2012o). *Qualcomm to add support for Java in M2M chipsets.* Accessed October 3, 2012, from http://www.cellular-news.com

Cellular-news. (2012p). *M2M market to reach 400 million units by 2017.* Accessed October 25, 2012, from http://www.cellular-news.com

Cellular-news. (2012q). *Ericsson signs M2M deal with Russia's MegaFon.* Accessed November 1, 2012, from http://www.cellular-news.com

Cellular-news. (2012r). *TeliaSonera extends M2M agreement with Ericsson.* Accessed November 6, 2012, from http://www.cellular-news.com

Cellular-news. (2012s). *Machine-to-machine connections growing to more than 325 million by 2016.* Accessed November 27, 2012, from http://www.cellular-news.com

Cellular-news. (2012t). *Jasper Wireless #1 in M2M – but watch out for Ericsson.* Accessed November 28, 2012, from http://www.cellular-news.com

Cellular-news. (2013a). *InterDigital forms M2M joint-venture with Sony.* Accessed January 3, 2013, from http://www.cellular-news.com

Cellular-news. (2013b). *Ericsson and Gemal to work on simplifying M2M device management.* Accessed February 11, 2013, from http://www.cellular-news.com

Cellular-news. (2013c). *Jasper Wireless signs up 2,000 customers to M2M services.* Accessed February 19, 2013, from http://www.cellular-news.com

Cellular-news. (2013d). *Ericsson and SAP announce new combination of cloud-based M2M solutions.* Accessed March 6, 2013, from http://www.cellular-news.com

Cellular-news. (2013e). *Mobile payment transactions may reach $1 trillion by 2015.* Accessed March 9, 2013, from http://www.cellular-news.com

Cellular-news. (2013f). *M2M will drive mobile operator data revenues past voice revenues by 2018.* Accessed March 9, 2013, from http://www.cellular-news.com

Cellular-news. (2013g). *NFC installed base to exceed 500m devices within 12 months.* Accessed March 26, 2013, from http://www.cellular-news.com

Cellular-news. (2013h). *Telecom operators are leaning toward M2M platform control.* Accessed March 30, 2013, from http://www.cellular-news.com

Cellular-news. (2013i). *Telenor Connexion to offer M2M services for KDDI customers.* Accessed May 2, 2013, from http://www.cellular-news.com

Cellular-news. (2013j). *Telefónica strikes Europe-wide M2M deal with Dell.* Accessed May 7, 2013, from http://www.cellular-news.com

Cellular-news. (2013k). *Wyless expands M2M contract with Telefonica.* Accessed May 21, 2013, from http://www.cellular-news.com

Cellular-news. (2013l). *The infrastructure for NFC mobile wallet service is being rolled out.* Accessed June 5, 2013, from http://www.cellular-news.com

Cellular-news. (2013m). *M2M connections to exceed 450 million by 2018 as ARPUs are squeezed.* Accessed July 29, 2013, from http://www.cellular-news.com

Coetzee, L., & Eksteen, J. (2011). *The internet of things – promise for the future? An introduction.* IIMC International Information Management Corporation. Accessed June 11, 2013, from http://www.ist-africa.org

Cristaldi, L., Faifer, M., Grande, F. et al. (2005, February 8–10). *An improved M2M platform for multi-sensor agent applications.* Paper presented at SIcon/05 – Sensors for Industry Conference, Houston, TX.

Cukier, K. (2010, February 27). Data, data everywhere – A special report in managing information. *The Economist.*

Cukier, K., & Mayer-Schoenberger, V. (2013, May/June). The rise of big data. *Foreign Affairs,* 28–40.

Curwen, P., & Whalley, J. (2010). *Mobile telecommunications in a high-speed world – industry structure, strategic behaviour and socio-economic impact.* Farnham: Gower.

Darianian, M., & Michael, M. (2008). *Smart home mobile RFID-based Internet-of-Things systems and services.* International Conference on Advanced Computer Theory and Engineering. Accessed June 11, 2013, from http://www.ieeexplore.ieee.org

Detecon Consulting. (2012). *Eliminating regulatory roadblocks for M2M deployment.* Accessed July 29, 2013, from http://www.detecon.com

Domingo, M. (2012). An overview of the internet of things for people with disabilities. *Journal of Network and Computer Applications, 35,* 584–596.

Dorasamy, M., Raman, M., & Kaliannan, M. (2013). Knowledge management systems in support of disasters management: A two decade review. *Telematics and Informatics,* forthcoming.

Doukas, C., Metsis, V., Becker, E., Le, Z., Makedon, F., & Maglogiannis, I. (2011). Digital cities of the future: Extending @home assistive technologies for the elderly and the disabled. *Telematics and Informatics, 28,* 176–190.

EE Times. (2011). *Cellular M2M modules predicted to see dramatic uptake in automotive and metering.* Accessed January 30, 2011, from http://www.eetimes.com

European Commission. (2008). *Internet of things in 2020 – a roadmap for the future.* Luxembourg: European Commission – Information Society and Media.

European Communications Committee. (2010, November). *Numbering and addressing in machine-to-machine (M2M) communications* (ECC Report 153). Luxembourg: Author.

Evans, P., & Annunziata, M. (2013). *Industrial internet: Pushing the boundaries of minds and machines.* Accessed July 26, 2013, from http://www.ge.com

Fadlullah, Z., Fouda, M., Kato, N., Takeuchi, A., Iwasaki, N., & Nozaki, Y. (2011, April). Toward intelligent machine-to-machine communications in smart grid. *IEEE Communications Magazine,* 60–65.

Fleisch, E. (2010). *What is the internet of things?* Auto-ID Labs White Paper WP-BIZAPP-053.

Foschini, L., Taleb, T., Corradi, A., & Bottazzi, D. (2011, November). M2M-based metropolitan platform for IMS-enabled road traffic management in IoT. *IEEE Communications Magazine,* 50–57.

Gang, G., Zeyong, L., Jun, J. (2011). *Internet of things security analysis. Internet Technology and Applications.* Accessed June 11, 2013, from http://www.ieeexplore.ieee.org

GSMA. (2008). *Embedded mobile – M2M solutions and beyond.* Accessed November 30, 2008, from http://www.gsma.org

GSMA. (2011). *Smart mobile cities: Opportunities for mobile operators to deliver intelligent cities.* Accessed December 12, 2011, from http://www.gsma.org

GSMA. (2013a). *Smart city resilience.* Accessed January 28, 2013, from http://www.gsma.org

GSMA. (2013b). *Guide to smart cities – the opportunity for mobile operators.* Accessed February 27, 2013, from http://www.gsma.org

Guinard, D., Trifa, V., Mattern, F., & Wilde, E. (2011). From the internet of things to the web of things: Resource-oriented architecture and best practices. In D. Uckelmann, M. Harrison, & F. Michahellas (Eds.), *Architecting the internet of things.* Heidelberg: Springer.

Haifei, Y. (2010, January). M2M: The Internet of 50 billion devices. *Win-Win Magazine,* 19–22.

Haller, S. (2010). *The things in the Internet of Things.* Accessed June 11, 2013, from http://www.lot-a.eu

Haller, S., Karnouskos, S., & Schroth, C. (2009). The internet of things in an enterprise context. In J. Domingue, D. Fensel, & P. Traverso (Eds.), *Future internet – first internet symposium.* Heidelberg: Springer.

Hu, R. Q., Qian, Y., Chen, H-H., & Jamalipour, A. (2011, April). Guest editorial – recent progress in machine-to-machine communications. *IEEE Communications Magazine,* 24–26.

International Telecommunication Union. (2005). *The internet of things – executive summary*. Switzerland: ITU.

Kim, D.-H., & Song, J. (2008). Mobile and remote operation for M2M application in upcoming u-manufacturing. *Journal of Mechanical Science and Technology, 22*, 12–24.

Kosmatos, E., Tselikas, N., & Boucouvals, A. (2011). Integrating RFIDs and smart objects into a united internet of things architecture. *Advances in Internet of Things, 1*, 5–12.

Lawton, G. (2004). Machine-to-machine technology gears up for growth. *IEEE Computer, 37*, 12–14.

Lee, J., Phaal, R., & Lee, S.-H. (2013). An integrated service-device-technology roadmap for smart city development. *Technological Forecasting and Social Change, 80*, 286–306.

Li, X., Lu, R., Liang, X., Shen, X., Chen, J., & Lin, X. (2011, November). Smart community: An internet of things applications. *IEEE Communications Magazine*, 68–75.

Lien, S-Y., Chen, K-C., Lin, Y. (2011, April). Toward ubiquitous masses assesses in 3GPP machine-to-machine communications. *IEEE Communications Magazine*, 66–74.

Liu, J., Li, X., Chen, X., Zhen, Y., & Zeng, L. (2011). *Applications of internet of things on smart grid in China. ICACT*. Accessed June 11, 2013, from http://www.ieeexplore.ieee.org

Lu, R., Li, X., Liang, X., Shen, X., & Lin, X. (2011, April). GRS: The green, reliability and security of emerging machine to machine communications. IEEE Communications Magazine, 28–35.

Mainetti, L., Patrono, L., Vilei, A. (2011). *Evolution of wireless senor networks towards the Internet of Things: A survey. Software, Telecommunications and Computer Networks*. Accessed June 11, 2013, from http://www.ieeexplore.ieee.org

Middleton, C., & Bryne, A. (2011). An exploitation of user-generated wireless broadband infrastructures in digital cities. *Telematics and Informatics, 28*, 163–175.

Miorandi, D., Sicari, S., de Pellegrini, F., et al. (2012). Internet of things: Vision, applications and research challenges. *Ad Hoc Networks, 10*, 1497–1516.

Morrish, J. (2012). *Mobile operator strategies for success in M2M: The role of cloud platforms. Machina Research. September*. Accessed July 10, 2013, from http://www.jasperwireless.com

Nique, M., Arab, F. (2013). *Sustainable energy and water access through M2M connectivity. GSMA Mobile Enabled Community Services*. Accessed May 10, 2013, from http://www.gsma.org

Niyato, D., Xiao, L., & Wang, P. (2011, April). Machine-to-machine communications for home energy management system in smart grid. *IEEE Communications Magazine*, 53–59.

OECD. (2012, January 30). *Machine-to-machine communications: Connecting billions of devices*. DSTI/ICCP/CISP(2011)/4/FINAL.

Pauchant, T., & Mitroff, I. (1990). Crisis management – managing paradox in a chaotic world. *Technological Forecasting and Social Change, 38*, 17–134.

Rezende, D.A., Madeira, G. dos S., Mendes, L., et al. (2013). Information and telecommunication project for a digital city: A Brazilian case study. *Telematics and Informatics*, forthcoming.

Siegele, L. (2010, November 6). It's a smart world – A special report on smart systems. *The Economist*.

Siorpaes, S., Broll, G., Paolucci, M., Rukzio, E. (2006). *Mobile interaction with the internet of things*. Po54ster. 4th International Conference on Pervasive Computing, Dublin, Ireland.

Sony Corporation of America. (2013). *Sony and interdigital team to launch machine-to-machine focused joint venture called Convida Wireless*. Accessed March 1, 2013, from http://www.sony.com/SCA

Standage, T. (2009, September 26). Mobile marvels – A special report on telecoms in emerging markets. *The Economist*.

Starsinic, M. (2010). *System Architecture Challenges in the Home M2M Network*. Applications and Technology Conference. Accessed June 11, 2013, from http://www.ieeexplore.ieee.org

Tan, S., Sorriyabandara, M., & Fan, Z. (2011). M2M communications in the smart grid: Applications, standards, enabling technologies and research challenges. *International Journal of Digital Multimedia Broadcasting*. Article ID: 289015.

Taylor, P. (2013). M2M: Telecom giants' data expansion hopes. Financial Times. Accessed May 22, 2013, from http://www.ft.com

Telecom.com. (2013). *MNOs will not make the IoT connection.* Accessed July 26, 2013, from http://www.telecoms.com

Vermesan, O., Harrison, M., Auto-ID Lab, et al. (2009). *Internet of Things – Strategic Research Roadmap.* 15 September. Accessed June 11, 2013, from http://www.internet-of-things-research.eu

Weber, R. H. (2009). Internet of things – need for a new legal environment? *Computer Law and Security Review, 45,* 522–527.

Wu, G., Talwar, S., Johnsson, K., Himayat N., & Johnson K.D. (2011, April). M2M: From mobile to embedded internet. *IEEE Communications Magazine.* 36–43.

Zhang, Y., Yu, R., Xie, S., Yao, W., Xiao, Y., & Guizani, M. (2011, April). Home M2M networks: Architectures, standards and QoS improvement. *IEEE Communications Magazine.* 44–52.

Zhao, J., Zheng, X., Dong, R., & Shao, G. (2013). The planning, construction, and management toward sustainable cities in China needs the environmental Internet of Things. *International Journal of Sustainable Development and World Ecology, 20*(3), 195–198.

Zorzi, M., Gluhak, A., Lange, S., Bassi, A. (2010, December). From today's Intranet of things to a future internet of things: A wireless and mobility related view. *IEEE Wireless Communications.* 44–51.

Index

P. Curwen and J. Whalley, *Fourth Generation Mobile Communication*,
Management for Professionals, DOI 10.1007/978-3-319-02210-9,
© Springer International Publishing Switzerland 2013